The Practical Bee Guide
A Manual of Modern Bee-keeping

by Rev. J.G. Digges

with an introduction by Jackson Chambers

This work contains material that was originally published in 1921.

This publication is within the Public Domain.

*This edition is reprinted for educational purposes
and in accordance with all applicable Federal Laws.*

Introduction Copyright 2018 by Jackson Chambers

COVER CREDITS

Front Cover
Silphium perfoliatum and Apis mellifera By Rillke (Own work)
[CC0 – Universal Public Domain Dedication],
via Wikimedia Commons

Back Cover
Lornado Honey Extraction by US Embassy Canada
[CC BY 2.0 - http://creativecommons.org/licenses/by/2.0],
via Wikimedia Commons

Research / Resources
Wikimedia Commons
www.Commons.Wikimedia.org

Many thanks to all the incredible photographers, artists,
researchers, and archivists who share their great work.

PLEASE NOTE :
As with all reprinted books of this age that are intended to perfectly reproduce the original edition, considerable pains and effort had to be undertaken to correct fading and sometimes outright damage to existing proofs of this title. At times, this task can be quite monumental, requiring an almost total rebuilding of some pages from digital proofs of multiple copies. Despite this, imperfections still sometimes exist in the final proof and may detract slightly from the visual appearance of the text.

DISCLAIMER :
Due to the age of this book, some methods or practices may have been deemed unsafe or unacceptable in the interim years. In utilizing the information herein, you do so at your own risk. We republish antiquarian books without judgment or revisionism, solely for their historical and cultural importance, and for educational purposes.

Self Reliance Books

Get more historic titles on animal and stock breeding, gardening and old fashioned skills by visiting us at:

http://selfreliancebooks.blogspot.com/

introduction

Here at **Self-Reliance Books** we are dedicated to bringing you the best in *dusty-old-book-knowledge* to help you in your quest for self-sufficiency and food independence.

We're so pleased to bring you this old title on Apiculture. Not only is honey sweet and divine-tasting, it also has healing properties, including antimicrobial and antiviral actions.

Raw, organic, locally-produced honey is also said to help alleviate the pain and suffering of seasonal allergy sufferers - a great item to have in your survival pantry when *Sneezing Season* begins!

This special edition of **The Practical Bee Guide : A Manual of Modern Bee-Keeping** was written by the Reverend J.G. Digges, and first published in 1921, making it just shy of a century old.

The book has sections on *The Occupants of the Hive, The Bee in Spring, The Bee in Summer, The Bee in Autumn and Winter, Anatomy of the Bee, Different Races of Bees, Bee Products, Comb Foundation,* and more.

A wonderful old tome filled with good, old-fashioned knowledge needed to begin a career in Apiculture, and a great place to start for those considering it.

~ *Roger Chambers*
State of Jefferson, March 2018

CONTENTS.

	PAGE
Preface to the First Edition	ix.
Preface to the Second Edition	vii.
Preface to the Third Edition	v.
Preface to the Fifth Edition	vi.
Note	x.

Chapter I.—The Occupants of the Hive ... 1
General Remarks, 1, 2. Occupants of the hive, 3. Queen, 4. Workers, 5. Drones, 6.

Chapter II.—The Bee in Spring ... 4
Signs of survival, 7. Breeding begins, 8. Work out of doors commences, 9. Wax production and comb building, 10. Sanitation in the hive, 11. Guarding the portal, 12. Approach of summer, 13.

Chapter III.—The Bee in Summer ... 7
A crisis, 14. The Mysterious Influence, 15. Indomitable spirits, 16. Queen rearing, 17. The swarm—a deliberated sacrifice, 18. The swarm—an ecstacy, 19. Virgin Queen, 20. Queen's wedding, 21. Parthenogenesis, 22. A splendid example, 23.

Chapter IV.—The Bee in Autumn and Winter ... 13
Death of the Drones, 24. Approach of winter, 25.

Chapter V.—Anatomy of the Bee ... 15
General Remarks, 26. External skeleton, 27. Head, 28. Simple eyes, 29. Compound eyes, 30. Antennæ, 31. Organs of mouth, 32. Thorax, 33. Legs, 34. Wings, 35. Spiracles and Tracheæ, 36. Abdomen, 37. Honey sac, 38. Worker's sting, 39. Palpi, 40. Queen's sting, 41. Organs of Drone, 42. Organs of Queen, 43. Parthenogenesis, 44. Fertilisation of egg, 45.

Chapter VI.—Different Races of Bees ... 30
Black or Native bees, 46. Italians or Ligurians, 47. Carniolans, 48. Cyprians, 49. Syrians, 50. Giant, 51. Common East Indian, 52. Dwarf East Indian, 53. Dutch, 54. Sand, 55. Leafcutter, 56. Caucasians, 56b.

Chapter VII.—Bee Products, &c. ... 33
Honey, 57. Gathering and storing honey, 58. Water in honey, 59. Honey as food, 60. Honey Dew, 61. Beeswax, 62. Honey used in wax production, 63. Paraffin wax and Ceresin wax, 64. Honey comb, 65. Worker cells, 66. Drone cells, 67. Hexagonal cells, 68. Transitional or Intermediate cells, 69. Use of cells for storing, 70. Queen cells, 71. Cappings, 72. Value of combs, 73. Pollen, 74. Propolis, 75. Adulteration of Honey, 75b.

CONTENTS.

PART II.—Hives and Appliances.

PAGE

Chapter VIII.—Hives and Frames 40
Ancient hives, 76. The skep, 77. Uses of skeps, 78. The skep giving place to the moveable-comb hive, 79. Genesis of the moveable-comb hive, 80. Advantages of the moveable-comb hive, 81. The hive in general use in Ireland, 82. Internal measurements, 83. "Federation" hive, 84. Floor-board, 85. Body-box, 86. Lift or Riser, 87. Roof, 88. "W.B.C." hive, 89. Observatory hive, 90. "I.B.A. 1909" hive, 91. "Hibernian" hive, 92. Dummy or Division Board, 93. Use of Dummy, 94. "Federation" dummy, 95. Sheet and quilts, 96. Frames, 97. Various sizes of frames, 98. "Claustral Detention Chamber," 98b.

Chapter IX.—Appliances for Supering 53
Supering, 99. Section, 100. Sections of various kinds, 101. Separator, 102. Section crate, 103. Divisional crate, 104. Observatory crate, 105. Follower, 106. Hanging crate, 107. Super box, 108. Excluder, 109.

Chapter X.—Comb Foundation 58
Use of, 110. Invention of, 111. Varieties of, 112. Advantages of, 113. Adulteration of, 114. Change of colour, 115. Quantity required, 116. Fixing foundation, 117. Wiring appliances, 118

Chapter XI.—Appliances for Feeding Bees 64
Feeding, 119. Economic feeder, 120. Bottle and stage feeder, 121. Graduated feeder, 122. Slow and rapid feeders, 123. Canadian feeder, 124. Division board feeder, 125.

Chapter XII.—Appliances for Subduing and Handling Bees 67
Smoker, 126. Carbolic cloth, 127. Use of veils, 128. Lady's veil, 129. Wire-cloth veil, 130. Use of gloves, 131. Various gloves, 132.

Chapter XIII.—Appliances for Honey and Wax Extraction 72
Invention of the honey extractor, 133. Honey extractor, 134. Uncapping knife, 135. Strainer and Ripener, 136. Honey press, 137. Wax extractors, 138. Solar wax extractor, 139. Steam wax extractor, 140.

PART III.—Modern Bee-Keeping.

Chapter XIV.—Past and Present 76
Past ignorance, 141. Survival of the unfit, 142. Modern bee-keeping, 143. A profitable industry, 144.

Chapter XV.—Arranging an Apiary 80
Selecting a position, 145. Bees near dwellings, 146. Position of the hives, 147. Appliance press and apiary house, 148.

CONTENTS.

PAGE

Chapter XVI.—Commencing Bee-Keeping 83
Three words of advice, 149. Begin on a small scale, 150. Purchasing bees, 151. Commencing with a swarm, 152. Moving swarms, 153. Sending swarms per post, 154. Commencing with a stock, 155. Moving stocks, 156. Moving stocks in skeps by road or rail, 157. Moving stocks in frame hives by road or rail, 158. Commencing with driven bees, 159. Driving bees, 160. Study the subject, 161.

Chapter XVII.—Subduing and Handling Bees ... 92
Tranquilising influence of smoke, 162. Unprovoked stinging exceptional, 163. Fearless defence of the home, 164. What constitutes a "Master of Bees," 165. Swarming bees—harmless, 166. Full of sweets—empty of bitterness, 167. A firm and gentle hand necessary, 168. Protection for beginners, 169. Treatment of stings, 170.

Chapter XVIII.—Manipulating 99
Appliances required, 171. Comb stand, 172. Comb box, 173. Vaseline and Petroleum jelly, 174. Preparing the smoker, 175. Preparing the carbolic cloth, 176. Opening the hive, 177. Manipulating wicked stocks, 178. Forcing the pace, 179. Smoking overdone, 180. No food—no subjugation, 181. Examining the combs: finding the queen, 182. The combs described, 183. Removing bees from combs, 184. Turning combs, 185. Searching for the Queen, 185b.

Chapter XIX.—Breeding 108
Breeding begins, 186. Congestion to be guarded against, 187. Drone-breeding queens, 188. Age of larvæ, 189. Worker brood, 190. New combs for breeding, 191. Stimulating in spring, 192. Spreading the brood, 193. Drone brood, 194. Controlling drone rearing, 195. Queen cells, 196. Nursing queen larvæ, 197. Wonderful effects of special nursing, 198. Queen brood, 199. Laying workers, 200. Removing laying workers, 201. Stimulating in autumn, 202. Breeding ceases, 203. Metamorphosis, etc., of bees. 204.

Chapter XX.—Swarming 118
Natural swarming, 205. Signs of swarming, 206. Delay of swarming, 207. The swarm, 208. Vagaries of swarms, 209. To encourage clustering, 210. Truant swarms, 211. Clipping queens' wings, 212. The parent stock, 213. Casts, 214. Hunger swarms, 215. Prevention of swarming, 216. Giving room, 217. Ventilation, 218. Limiting drone rearing, 219. Limiting queen rearing, 220. Prevention of casts, 221. Artificial swarming, 222. Conditions, 223. One swarm from one colony, 224. One stronger swarm from two colonies, 225. Using three or more stocks. 226. Making swarms for sale, 227. One swarm from a stock and a nucleus, 228. Making swarms from stocks in skeps, 229. A stronger swarm from two stocks in skeps, 230

CONTENTS.

Chapter XXI.—Hiving: Uniting and Transferring Bees ... 130
Confidence in protection from stings, 231. Preparing the hive, 232. Hiving swarms direct, 233. Swarms in high trees, 234. Swarms in awkward places, 235. Hiving from a skep, 236. Secure all the cluster, 237. Sweetening the hiving skep, 238. Hiving by caging, 239. Hiving a swarm on the old stand, 240. Heddon method, 241. Returning swarms, 242. Retracing swarms, 243. Uniting bees, 244. Uniting swarms, 245. Uniting two stocks, 246. Uniting queenless bees to a stock, 247. Uniting a swarm to a stock, 248. Uniting driven bees, 249. Uniting driven bees to a stock, 250. Transferring bees, 251. Transferring from hive to hive, 252. Transferring from skep to modern hive, 253. Automatic transfer from skep, or box, to modern hive, 254. Heddon method of transfer, 254b. Separating swarms, 254c.

Chapter XXII.—Surplus Honey 139
Preparing in time, 255. Extracted honey more profitable than comb honey, 256. Preparing crates and sections, 257. Three-split sections, 258. Split top sections, 259. Unsplit sections, 260. Preparing frames, 261. Wiring frames, 262. Fixing foundation in frames, 263. Three "Donts," 264. The honey flow, 265. Putting on crates, 266. Putting on super boxes, 267. Use of excluders, 268. Tiering crates, 269. Doubling and storifying, 270. Supering skeps, 271. Removing supers, 272. Use of cone escapes, 273. Super clearer, 274. Use of super clearer, 275.

Chapter XXIII.—Extracting Honey 154
Extracting, 276. Straining and ripening, 277. Cleaning extracted combs, 278.

Chapter XXIV.—Extracting Wax 157
Use of wax extractors, 279. Extracting by boiling, 280.

Chapter XXV.—Queen Rearing and Introduction ... 158
Old queens, 281. Defective queens, 282. Queenlessness, 283. Signs of queenlessness, 284. Nucleus hives, 285. Queen rearing, 286. Using a swarmed stock, 287. Returned swarm method, 288. Using an unswarmed stock, 289. Forming nuclei, 290. Inserting queen cells, 291. Management of nuclei, 292. Using two stocks, 293. Distributing the nuclei, 294. Queen introduction, 295. Balling the queen, 296. Use of queen cages, 297. Introduction by artificial swarming, 298. Direct introduction, 299. Sending queens per post, 300. Queen rearing on a large scale, 300b.

Chapter XXVI.—Marketing Honey 169
Home honey, 301. Storing honey, 302. Preparing comb honey for market, 303. Glazing sections, 304. Packing sections for transport, 305. Preparing and packing extracted honey for market, 306

Chapter XXVII.—Robbing and Fighting 175
Robbing, 307. Precautions, 308. Signs of robbing, 309. Treatment, 310.

CONTENTS.

Chapter XXVIII.—Feeding Bees: Recipes 178
Objects of feeding bees, 311. Precautions, 312. Spring feeding, 313. Summer feeding, 314. Autumn feeding, 315. Winter feeding, 316. Feeding for comb building, 317. Feeding bees in skeps, 318. Water, 319. Pollen, 320. Recipes—Spring and summer syrup, 321. Autumn syrup, 322. Candy for Winter food, 323. Flour candy, 324. Naphthol Beta solution, 325. Measures, 326. Syrup from Candy, 326b.

Chapter XXIX.—Diseases, &c. 185
Diseases, 327. *Dysentery*, 328. Symptoms, 329. Cause, 330. Prevention, 331. Treatment, 332. *Paralysis*, 333. Symptoms, 334. Treatment, 335. *Chilled brood*, 336. Symptoms, 337. Cause, 338. Prevention, 339. Treatment, 340. *Black brood*, 341. Symptoms, 342. Cause, 343. Treatment, 344. *Pickled brood*, 345. Symptoms, 346. Cause, 347. Treatment, 348. *Foul brood*, 349. Symptoms, 350. Cause, 351. Prevention, 352. Treatment, 353. Early stages—treatment with formalin, 354. Advanced stages—treatment by burning, 355. Treatment by artificial swarming, 356. Requeening desirable, 357. Infected honey dangerous, 358. Disinfecting necessary, 359. "American" and "European" Foul Brood, 359b. "Isle of Wight Disease," 360. Differential diagnosis, 361. Recipes—Carbolic solution for subduing bees, 362. Carbolic solution for disinfecting hives, 363. Carbolic solution for disinfecting clothing, etc., 364. Formalin solution for injecting into diseased cells, 365. Formalin solution for use under combs, 366. Izal solutions, 363, 364, 365, 366.

Chapter XXX.—Enemies of Bees 202
Enemies, 367. Ants, 368. Birds, 369. Earwigs, 370. Mice, 371. Parasites, 372. Wasps, 373. Wax moth, 374.

Chapter XXXI.—Wintering 205
Successful wintering, 375. Strong stocks, 376. Sufficient wholesome food, 377. Quiet, 378. Ventilation, 379. Damp and storms, 380.

Chapter XXXII.—Work for the Month 208
January to December, 381,—392.

Chapter XXXIII.—Exhibiting and Judging Bee Products 210
Points to be aimed at, 393. Early exhibition sections, 394. Mid-season ditto, 395. Heather ditto, 396. Selecting ditto, 397. Preparing ditto, 398. Extracted Clover, or Light, Honey for Exhibition, 399. Extracting and Preparing ditto, 400. Extracted Heather or Dark Honey for Exhibition, 401. Extracting and Preparing ditto, 402. Supers of Honey for Exhibition, 403. Beeswax for Exhibition, 404. Mead for Exhibition, 405. Vinegar for Exhibition, 406. Judging Bee Products, 407.

Chapter XXXIV.—Bee Flowers and Plants 221

Index 224

PREFACE TO THE FIFTH EDITION.

This edition brings the number issued up to 25,000. It has been revised throughout. New matter has been added, especially in Chapters XXV, "Queen Rearing," and XXIX, "Diseases, Etc."; and a number of additional illustrations have been inserted. The popularity of the Guide increases yearly, and is evident by the rapidity of sales, as well as by most gratifying correspondence received from a large number of readers since the fourth edition appeared, in 1919.

J. G. DIGGES.

Clooncahir, Lough Rynn,
    Co. Leitrim, July 1, 1921.

PREFACE TO THE THIRD EDITION.

The popularity of this Guide has been well established, and a third edition has been called for. This new edition brings the number issued up to 13,000. It has been enlarged and improved by the addition of fresh matter and the introduction of new illustrations. I have again to express my sincere thanks to many hundreds of correspondents whose generous approval of the Guide has been to me most welcome and encouraging.

J. G. DIGGES.

Clooncahir, Lough Rynn,
    Co. Leitrim, May 23rd, 1917.

PREFACE TO THE SECOND EDITION.

The generous reception which was accorded to the first edition of this GUIDE more than justified the publication of the book, and confirmed the opinion that "there was need for a guide to Beekeeping which should supply information and advice of a more extensive nature than any yet published in this country." The Press reviews, without an exception, commended it, not only in the United Kingdom, but also in the Colonies and in far away foreign lands. The craft—practical beekeepers engaged in the industry and capable of judging by experience—welcomed the book, and a large number of these were kind enough to write to me expressing their approval in very gracious words. Several hundreds of such letters were received, and welcomed as evidences of that good nature in bee-lovers which has become proverbial the world over and has placed me under obligations to many whom, otherwise unknown, a mutual interest has constituted familiar and faithful friends. Not a few of those communications were such as might well compensate any man for years of investigation and work. "I have read it at meals, read it at night, and read it at dawn, and from a woman simply desirous of earning a little money by keeping bees, I have become an enthusiast": an English correspondent put it so. From an earnest, devoted monk, in another country, came the words—"I never fail to carry it with me as a good companion whenever I am absent from home." From Australia, a practical apiarist wrote—"I travel all over the State as Government Expert, but never without the GUIDE. I have read and re-read it. It has fascinated me. It is like The Old Book—always interesting." From a 40-foot canoe, on the river of Uganda, a travelling official wrote in similar strains. The book has reached the most distant parts, and there, it is hoped, as well as in these countries, has achieved some, at least, of the objects with which it was published.

For the present edition the original work has been thoroughly revised. Many new paragraphs have been added, treating of such subjects as the "W.B.C." Hive, the "I.B.A. 1909" Hive, "Claustral Detention Chambers," "Searching for the Queen," the "Isle of Wight Disease," Recent Investigations into the Cause of Foul Brood, etc., and a new chapter on "Exhibiting and Judging Bee Products" has been introduced in response to a frequently expressed wish. The number of illustrations has also been increased by the insertion of 20 new blocks, while, of those in the first edition, I have removed

PREFACE.

53, supplying their places with others, more accurate, deeming it wise, if not, indeed, necessary, to rely upon my pen and camera for the illustration of manipulations, appliances, etc., which, usually prepared by a cheaper process, are not always so satisfactorily presented. In these respects this Second Edition will be found to be a distinct improvement upon the first.

I gratefully acknowledge my indebtedness to the works of numerous writers who have preceded me in such investigations, from whose teaching I have learned much in a study covering a period of nearly a quarter of a century. Among these must be mentioned the well-known works of Mr. T. W. Cowan— "The Honey Bee," and "The British Bee-keepers' Guide Book," (now in its nineteenth edition), from the sixth edition of which (in 1885) I learned my first lessons in beekeeping; "Bees and Beekeeping," by the late F. Cheshire, a classic, now out of print, from which the publisher, Mr. Upcott Gill, in addition to supplying the blocks enumerated in the Note, page x., has generously permitted me to supply my readers with much valuable information; and, besides these, a host of writers of whose works I have made an exhaustive study, including Bagster, Bevan, Cotton, Dzierzon, De Galieu, Huber, Huish, Hutchinson, Hunter. J., (*Phil. Trans.*), Hunter. John, Hyatt, Keys, Kirby and Spence, Langstroth, Leukart, Lubbock, Miller, Milton, Miner, Neighbour, Nutt, Packard, Payne, Pettigrew, Pettitt, Pratt, Reaumur, Reid, Richardson, Root, Samuelson, Siebold, Simmins, Smith, Taylor, Thorley, Warder, Wighton, Wildman, Wood, and many others—some of these long out of print, but not one of them from which a diligent student may not learn something. In the preparation of Chapter XXXIII. I had the valuable assistance of Mr. M. H. Read, Hon. Secretary, Irish Beekeepers' Association, whose experience as a successful exhibitor and judge, was unreservedly placed at my disposal.

The alteration in the title of the GUIDE has been made partly in acknowledgment of the fact that the sale of the book hitherto has not been chiefly in this country, and partly in deference to the wishes of the booksellers and of a large number of practical beekeepers, who have assured me that the former title led to the erroneous impression that the GUIDE was suited only to beekeeping in Ireland. Many new titles were suggested to me: I have adopted one which appears to me to be not extravagant, for, whatever shortcomings the GUIDE may disclose, I think that I may, without immodesty, claim for it that it is essentially practical.

I offer my most sincere thanks to all who have encouraged me by their approval and patronage of a work the sale of

PREFACE.

which has far exceeded my expectations, and I issue this revised, enlarged, and improved edition in the hope that it may enjoy a like popularity, and may prove to be a reliable GUIDE for such as are interested, or may become interested, in the fascinating and profitable industry of Beekeeping.

J. G. DIGGES.

Clooncahir, Lough Rynn,
    Co. Leitrim, May 23rd, 1910.

PREFACE TO THE FIRST EDITION.

Queries, numbering several hundred, referred to me as Editor of the *Irish Bee Journal* during the last three years, have convinced me that there is need for a Guide to Bee-Keeping which shall supply information and advice of a more extensive nature than any yet published in this country, and in fuller detail than could be accommodated in the columns of a newspaper or periodical. Accordingly, and at the request of several prominent Bee-Keepers interested in the spread of the Industry, I have written this IRISH BEE GUIDE, in the hope that it may help to promote a wider knowledge of the wonders of Bee life, to encourage humane and intelligent treatment of the Honey Bee, and to assist the development of Bee-Keeping in Ireland as a National Industry.

J. G. DIGGES.

Clooncahir, Lough Rynn,
    Co. Leitrim, May 23rd, 1904.

NOTE.—This GUIDE consists of Three Parts. Part i. (pp. 1-39), deals with the History and Anatomy of the Bee, and with Bee Products: Part ii. (pp. 40-75), describes the Hives and Appliances generally in use: Part iii. (pp. 76-223), consists of Practical Directions for Management, with instruction for exhibitors and judges of Bee Products and a concluding chapter on Bee Flowers and Plants. The GUIDE is arranged in numbered and titled paragraphs. Where, in any paragraph reference to subjects dealt with in other portions of the book is desirable, the paragraph numbers are inserted in brackets, thus obviating the necessity for frequent examination of the Index, and facilitating reference to the subjects required. Of the 149 illustrations in the GUIDE, 125 are from original photographs by the author, and pen-and-ink sketches drawn specially for this work. The author gratefully acknowledges his indebtedness to the following for permission to publish the illustrations notified after their names:—Mr. W. Z. Hutchinson, Flint, Mich., U.S.A., Fig. 2. Mr. L Upcott Gill, London, Figs. 3, 4, 5, 6, 7, 8, 9, 10, 11, 12, 14. 78, 109. The A. I. Root Co., Medina, Ohio, U.S.A., Fig. 50. The Irish Bee Journal, Ltd, Figs 74. 112. and illustrations on pages 52b, 99, and 150.

THE PRACTICAL BEE GUIDE.

PART I.

THE HONEY BEE.

CHAPTER I.

THE OCCUPANTS OF THE HIVE.

Kingdom—Animal. *Sub-Kingdom*—Annulosa. *Division*—Arthropoda. *Class*—Insecta. *Order*—Hymenoptera. *Family*—Apidæ. *Genus*—Apis. *Species*—Mellifica.

> "Therefore doth heaven divide
> The state of man in divers functions,
> Setting endeavour in continual motion;
> To which is fixed, as an aim or butt,
> Obedience; for so work the honey-bees;
> Creatures that by rule in nature, teach
> The act of order to a peopled kingdom."
> — SHAKESPEARE.

1. It is natural that a guide to bee-keeping should begin with a description of the bees that are to be kept. And it is very necessary that everyone who desires to derive either pleasure or profit from the keeping of bees should know something of the bees which he proposes to keep—of their habits, their requirements, of the laws which govern their actions, and of the objects to which their marvellous energies and intelligence are devoted.

2. Therefore this guide begins with a description of the occupants of the hive, namely, the Queen, the Workers, and the Drones.

3. The Occupants of the Hive. In the summer months the hive of a prosperous colony of bees will be found, upon examination, to contain a queen, from 30,000 to 60,000 workers, and from 300 to 400 drones.

QUEEN.   WORKER.   DRONE
Fig. 1.

Photo by J. G. Digges.

4. The Queen (Fig. 1) is not the sovereign ruler of the bee kingdom, as her name might imply. She is neither daughter, wife nor widow of a king. She is obedient rather than commanding; and yet a queen in her own right; born to the purple; pre-eminent and distinguished above all others; the abundant mother, carrying in her prolific womb the creation and hope of unnumbered millions of her race. Hers is the longest life, extending to several years. Her very movements are queenly, the stately pace among her children marking her out to the observant as distinct from other occupants of the hive. In size, and form, and colour she is unique; longer, more delicately moulded, darker in hue. Her mission is to propagate; and for that most holy office nature endows her richly. Mated once for all, her strength, her life to it are unceasingly devoted. Within the hours that make a day and night 3,000 eggs from her teeming flanks may fall; and this prodigious labour will cease only with exhaustion of fecundity or approach of death. **(45).**

5. The Workers (Fig. 1) are the smallest bees in the colony; females, like the queen, but undeveloped. Theirs is a brief life, full of toil, of work so incessant that in the full flow of summer activity it yields to the pressure of exacting duty; and within a few weeks they drop and die, sacrificed to the demands of destiny, martyrs to the common good. If born in the autumn months they can survive the winter time of rest, and with the opening spring begin the work which unborn generations are to take up and carry to completion. Their responsi-

---

The figures in brackets, thus **(45)**, refer to the paragraphs bearing the numbers indicated.

bility is exceeding great; their labour is magnificent. They are the gatherers who, when nature decks the country side with fresh beauties, sally forth, and hurrying ever from flower to flower, collect the nectar, and pollen with which to feed the young, and propolis to fill up cracks and make the hive more homely. They manufacture wax, and with it build the combs which serve as cradles of the race, and larders for the store of honey. They feed the queen, nurse the young, cleanse the hive, and set up portal-guards to defend from all aggression the citadel that holds the secret of their destiny—the treasure of their faithful hearts. Fearless, surpassing diligent, beautifully unselfish, their marvellous intelligence fits them for that stupendous enterprise to which their lives are devoted, and for which they gladly die. (15).

6. **The Drones** (Fig. 1) or male bees, are thick and bulky, not so long as the queen, but longer than the workers. These are the oft maligned noisy, buzzing bees—

"The lazy yawning drone"

of Shakespeare, and the harmless, innocent butts for the gibes of modern critics. Theirs is a life of brief dependence and submission. They gather no stores: nature has not fitted them to do so. The one object of their existence is to fertilise the young queens. To that end they are born, are tolerated in the colony, and are allowed free access to the honey cells. Theirs, also, is the sacrifice of life to duty; and such of them as survive to the close of autumn are driven out of the hive to end, in cold and hunger, a life which, if seemingly idle or useless, was, at least, inoffensive, and full of possibilities whose vastness fills with awe and amazement every thinking mind. (43).

CHAPTER II.

THE BEE IN SPRING.

7. Signs of Survival.—With the lengthening of the days, the living mass clinging to the hive-combs feels the quickening breath of spring, and the bees of the cluster begin to move. Those on the outside pass in to the warmer centre of the sphere. The sun, in genial humour peeping through the open door, gives to the long-imprisoned inmates assurance of kindlier conditions without; and the bee-man, watching for signs of survival, delights to see first one, and then another, and presently many of his little pets appear upon the alighting board. Discreet in their new-found joy, they risk no long excursion, nor venture over much. Scenting the freshness of the air, they seem to revel in it, and in the heat and light which stir the life in them. They move about the entrance; examine the doors and porch; meet and salute each other; and rising, fly for a moment in front of the hive. A gladsome hour this for the bee-man also; an infectious happiness. He knows now that snow and storms, and all the frost and cruel winter hardships have failed to work their devastation within the little home which his foresight and loving care secured and sheltered before the falling leaves had left the branches bare. With each succeeding sun the bees in larger numbers move abroad—creatures "fanatically cleanly," who will suffer much and long and yet refuse to sully the purity that their incessant care preserves within the hive. **(329)**

8. Breeding begins.—In this, the new year's opening month, begins that wondrous work on which the thoughts, and energies, and hopes of all the colony are concentrated **(186).** The queen, stirring in the centre of the cluster, communicates to all around her that the hour has come for which, through the long months of winter, they have lived and waited; and activity spreads throughout the hive. From cell to cell, within a small circle, she passes, examining each, and depositing therein a tiny egg. Upon it nurse bees will lavish most tender care. During three days they will hatch it; and then, the grub appearing, it shall be fed for five days with food of the sweetest and purest—honey and pollen drawn from the flowers in the previous summer and stored for this same purpose in adjacent combs. Then shall the cell be sealed, still warmed by the clustering nurses, until the larva, transformed into a nymph shall, one week later, emerge a perfect bee to share the labours and to participate in the busy, and often hazardous enterprises of the colony. **(204).**

9. Work Out of Doors commences.—Meanwhile the queen has enlarged the circles of her brood, and has ventured upon fresh combs. Her downy progeny are bursting their cells on every side; the population is increasing, and the temperature of the hive rises rapidly. Outside, a spirit of resurrection has entered into nature, in whose scenes of progressive loveliness everything that moves experiences a new joy.

> "The softly warbled song
> Comes from the pleasant woods, and coloured wings
> Glance quick in the bright sun that moves along
> The forest openings."

Advancing spring has rescued from the embrace of winter the purple anemone and yellow crocus, fresh as the morning dew, and lovelier than the robe of Solomon in the days of his glory; gorse has made the hill-sides golden; hazel, and silex, and dandelion open their attractions around the fields. And from out the hive come the busy workers to gather in the stores kind nature has provided, and in turn, to render her good offices by transfer of the fertilising dust from flower to flower **(74).** Where nectar is, they sip it; where pollen, their feathery hairs collect it, and in the little baskets (*corbiculæ*) with which their hindmost legs are furnished **(34)** they bear it home to feed the larvæ. Water also they will find, for breeding cannot progress without it; and propolis to fasten joints and to exclude unwelcome draughts **(75).** These safely delivered up to those who work within, they start afresh, nor cease their eager gathering until the fading light, or cooling atmosphere warns them that the life required to-morrow must not be sacrificed to-day.

10. Wax Production and Comb Building.—Within the hive there is proceeding a work most truly marvellous. Those bees whose part it is to supply material for the building of the combs, have fed themselves from stores of honey, and, clinging one to the other in shape of festoons first, to thus facilitate the climbing of the rest, have formed in compact cluster **(62).** There, motionless, during many hours they hang, retaining and increasing the heat within the mass until a high temperature is attained; when upon the ventral plates, or pockets, under the abdomen appear clear scales of wax **(37).** First transferring these to the mouth for preparation, they hand them over to the builders, who, taking them in their mandibles, construct with them the comb—the masterpiece "that touches absolute perfection," by which the bees have taught a lesson to the highest human intelligence, and have applied the shape and form which give the greatest capacity and strength with least expenditure of material, time, and labour. **(68)**.

11. Sanitation in the Hive.—Other bees fulfil a lowlier task and undertake the cleansing of the hive. The winter's dead they carry out for burial. The brood which, immature and chilled and lifeless, occupy cells that missed the cluster's nursing warmth, are seized and dragged away to safer sepulture lest they infect the living, and render unavailing the anxious labours of the colony. The floor board, littered with particles of broken comb, and pollen pellets, and dust from two hundred thousand tiny, restless feet that come and go unceasingly, is swept and cleaned. For, nothing that can be moved or torn asunder, and that is not sweet and pure like bees themselves and like the largess of the open flowers, may linger long among those cheerful toilers who, if cleanliness be next to godliness, are, of all the insect class, nearest heaven.

12. Guarding the Portal.—Others still, placed about the portal, keep guard upon the treasury. Their watchful office is to see that all who seek an entrance have lawful business there. These are the sleepless sentinels, well armed, who pounce at once upon stranger bees and drive them off; or with their poison-stings make execution upon such as, intent on robbery, are bold enough to risk a conflict. **(309)**.

13. Approach of Summer.—And the patient, earnest queen—a slave to duty and willing minister of all, encouraged by the steady flow of honey, puts forth her best endeavours. Comb after comb is filled from top to base with honey sealed, and hatching brood, and larvæ pearly white, and eggs like bits of silken thread upon the bases of the cells. Beneath the porch two ceaseless streams of merry bees pass and return. For currant, thorn, and sycamore have hurried into bloom, and summer, with its happy song and gladsome days, is near at hand.

> "Fresh flow'rs shall fringe the wild brink of the stream,
> And with the songs of joyance and of hope
> The hedgerows shall ring loud."

CHAPTER III.

THE BEE IN SUMMER.

14. A Crisis.—About the time that sees the clover showing white in growing meadows, affairs within the hive approach a crisis. 50,000 gatherers, speeding upon the fragrant breezes through every sunny hour of May, have carried home great quantities of nectar to fill to overflowing each vacant cell. The queen, who, possessed of an insatiable desire for re-production and in the full flow of maternal vigour, has increased by thousands daily the number of her children, now finds herself encroached upon in her domain. The combs are fully occupied. The hive is crowded. The little bands of "fanners" at the door exhaust themselves in vain endeavours to ventilate their over-heated home **(59).** The bees returning from the fields loiter at the entrance, and hesitate to add their presence to the close-packed mass within. Some will cluster there, victims of a strange inertia;

"The slow hours measuring off an idle day."

Within a week the hatching brood will add a new congestion. Plainly a crisis has arrived. Something must be done, and done at once; for in bee life, except in winter, inactivity is the extreme vice that merits naught of mercy.

15. The Mysterious Influence.—Now that subtle, mysterious Influence which governs the whole life of the bee from the moment in which she struggles from her uncapped cell, a downy, awkward infant, until worn out with strain of excessive industry she drops from some pink heather bell, in the autumn evening, to rise no more: that silent, persistent, irresistible Influence which orders the economy of the hive; inspires each tiny occupant with courage of a hero; makes all instinct with uniformity of splendid purpose; and endows them with glorious spirit of self-sacrifice above all human imitation—a willingness to leave all, to lose all, and to bear all that may be, for love of the race and reverence for its destiny—asserts itself. A tremor passes through the bees, and an entirely new emotion seizes them. That love of others which recks not of personal suffering; that awe of the future which counts not of present perils; that infrequent exaltation which beautifies self-abnegation, idealizes the Unknown, and yields up life itself for others—possesses them. Their patient, untiring labours have secured for them supreme success: now they will forfeit all. They have reached the highest point of affluence: now they will renounce their wealth and fall to poverty. Their home is furnished through, and stored with food abundantly: now they will leave to others the fruits their energies have borne, forsake their home, and rush out, wildly exuberant of happiness, to build again their fortune, or in houseless cold and hunger to die.

16. Indomitable Spirits.—Not, however, without their queen. She shall accompany them. It is not meet that they should too far court disaster **(209)**. Without her, they must inevitably perish. With her, they shall die indeed, yet live again in their successors. Nor will their indomitable spirits contemplate extinction. Let but some friendly nook be found—some cavity in a spreading tree whose advancing age provides a cradle for a new born race; there combs will form again, and eggs be tended, and every passer-by shall hear the humming music of the bees, down by the river side where

> " the curling waves
> That break against the shore, shall lull the mind
> By one soft impulse saved from vacancy."

17. Queen Rearing.—However, one all-important preparation must first be made. The bees which stay behind to nurse the growing brood must have a queen to raise the colony to strength again when the enthusiastic swarmers shall have carried off the venerated mother bee. This, by one stupendous miracle of nature, shall be accomplished. Not one, but many queens shall be provided, lest any untoward accident should mar the great design. The workers, eager to enter on their new adventure, construct some special cells **(196)** by sacrifice of other cells around them—cells larger and with thicker walls. In them the queen, with that sublime indifference to persona! advantage which at the moment actuates her, deposits eggs. These, which in ordinary course of nature would produce but worker bees—females undeveloped, incapable of impregnation—shall be supplied with richer food, and in more abundance; shall have their cells enlarged yet more, and strengthened, and made to hang, in shape like acorns, between the combs (Fig. 2); until the cells are capped, and the royal princesses are left to spin their silken veils and, within a week, to emerge as perfect virgin queens.

18. The Swarm—A Deliberated Sacrifice.—Meanwhile restlessness seizes the old queen, who sees that the fulfilment of her maternal duty has been applied to raise, within the kingdom which she alone has peopled, rival claimants to her throne. She is not satisfied. She hurries from comb to comb, vainly endeavouring to assert an authority long subordinated to the requirements of her children. She even threatens the young princesses in their waxen nurseries. Wild excitement results among the little citizens. The palace of peace and home of steady labour is thrown into confusion. It is all so novel, this mad disorder and revolution of which no drone or worker has had experience previously. It is the perplexing acceleration of deliberated sacrifice, coming suddenly, rushing headlong, like the bursting of a mountain torrent that cannot by any means be stayed. The vats of honey are opened, and multitudes are feeding eagerly; for suspected danger always leads the bees to lay in store for quick emergencies (167). The temperature has risen to a point insufferable. The queen and all her people realize that the moment has arrived for the inevitable, reckless sacrifice which, in its ready willingness to give up all for the future of the race, invests the swarm with that uncommon glory which, during long ages, has been recognised and admired by astonished man.

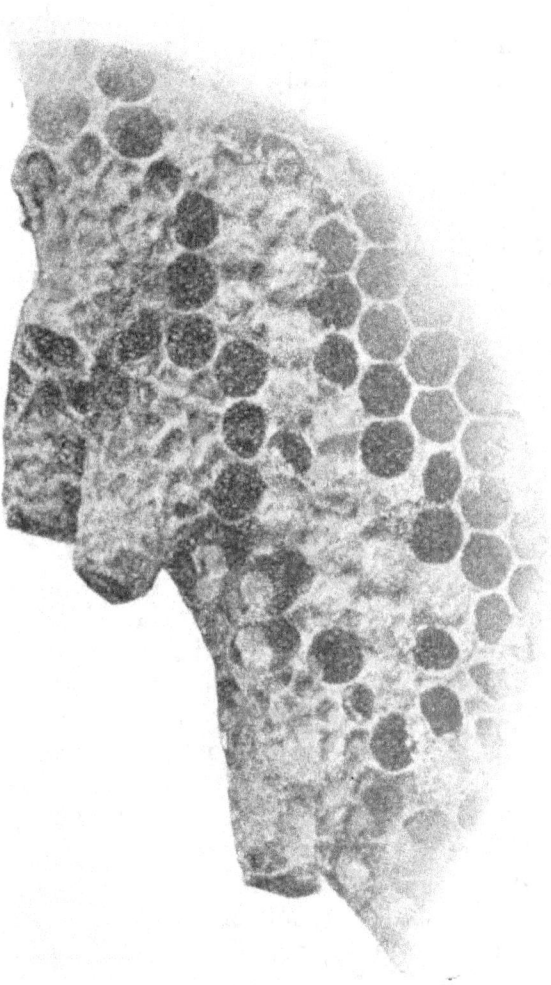

Fig. 2.
QUEEN CELLS.

19. The Swarm—An Ecstacy.—Pouring from the insufficient opening they come, in bewildering haste; a riotous throng, rapturously jubilant, in the very ecstacy of extravagant emotion; harmless, too, in their design, and in their exaltation so sweetly amiable that he who will may handle them in safety **(166).** A vast multitude it is, rushing hither and thither, with great noise of humming, until the queen has joined them from the hive and has alighted upon some neighbouring tree. Then they gather round her—in very numbers assuring her timid heart, unaccustomed to rough exposure and risk of outer dangers—and form a cluster with the faithful mother, so still that any passing traveller may hardly notice them. Now let the watchful owner hive them without delay, and set them to work in a new home, or they will rise and, following their scouts sent out before to find a dwelling, will settle in some distant tree or chimney, or will invade the ruined tower upon the neighbouring hill, and so be lost to useful purpose. **(208).**

20. The Virgin Queen.—The now depleted stock, deprived of more than half its numbers by that most boisterous exodus which robbed it also of its queen, presents once more a scene of peaceful labour. As if no mighty revolution had just disturbed their order, the bees pursue their avocations, apparently oblivious of the strange events which, but an hour ago, had shaken their kingdom to its foundations. A few days later the strongest of the young princesses is heard piping in her cell, as if conscious of the high importance of the position that awaits her, and impatient to attain it before her hatching rivals can intervene. The apex of her cell the workers have thinned and smoothed in order to assist her exit. Presently she will cut the capping and, pressing against it, force it open like a round, hinged lid (Fig. 14, A), and step out upon the comb. The nearest honey cell shall have her first attention; and then she who shall give life to unnumbered millions, will devote her first active hours to massacre. Reaching the other queen cells she will endeavour to tear them open at the sides and to slay her rivals **(199).** If this be not permitted, she will stay, and watch her opportunity to wage a battle-royal with any young princess who ventures abroad among the combs; or she will join an after-swarm **(214),** thus abdicating the position which, for so short a time and anxious, she occupied, and seeking peace in some new home where she may fulfil her task unhindered. But if the hive economy require no further division of the forces, the royal cells will be attacked (Fig. 14, B), and the occupants, astounded at this violent assault upon their privacy, be destroyed. "One queen, one kingdom," is, as in the domain of man, a law of bee life admitting few exceptions.

21. The Queen's Wedding.—So, in fifteen days from the depositing of the egg, a virgin queen has opened her astonished eyes upon the hive which is to be her home; upon the restless workers who come and go, and hurry back well laden; upon the drones, those bulky, strong-winged males whose lives, though short and helpless, are not devoid of joy; upon the combs that hold the nectar stores, and gilt-capped cells of hatching nymphs whose vacant places, when they emerge, she must occupy with living germs that shall produce a multitude, renewing month by month the population wasted by excessive toil. But this, not yet. So far she moves about unnoticed, in constant exploration that knows no instant's rest, and preparation for that wondrous incident which shall entitle her to claim the homage of her people, and to her queenly title add the higher, and more sacred name of "Mother." So far she has not felt the glow of sunshine, nor filled her tracheæ **(36)** with the breath of heaven. The eventful hour has not arrived. She must wait a few days more before she stakes herself, and all the secret of the future, upon the hazard of a flight. Then she approaches the entrance, inspecting everything, but not daring to venture farther. Again she appears, and hurries up and down; excited; impelled by that mysterious exaltation which nature pours out lavishly when great ends are to be accomplished by perilous enterprises. She spreads her wings and rises, quickly noting every little thing that marks the outworks of her citadel, and far more careful in this precaution than drone or worker, because of her exceeding value who carries in her person the hope and destiny of all. Pursuing wider circles she surveys the site until its every feature becomes familiar. Meanwhile upon the open flowers around, or resting on the sunny leaves, are countless drones, observing, each with his magnificent eyes of 26,000 hexagonal lenses **(30)**, the timid virgin's movements. Soon the loud humming of the full-fed males attracts the young queen, and as she enlarges the circles of her flight and passes over them, instantly they are in full pursuit. Here may be observed wise Nature's regulation that gives the battle to the strong, and to the brave the fair. The agile lover; he whose self-restraint has dipped with temperate appetite into the honey vats, and whose quick power of flight, not lessened by emasculating idleness, is trained and strengthened by sufficient exercise, is first to reach the queen, and in brief ecstacy of that embrace gives all his vigour to the making of a hardy race; and, giving all, he dies. **(42)**.

22. Parthenogenesis.—Thus mated once for all, the queen returns and meets a welcome from her people. Never will she leave the hive again, unless the swarming of the colony compels her. She will take up the task of supplying the vacant cells with eggs. Generation after generation shall live and die, and leave her still fulfilling her calling. Nor will several years exhaust the 25,000,000 spermatozoa which one short intercourse supplied **(43)**. Just here is disclosed another marvellous feature in the life of the bee. The drone which fertilised the queen; himself fatherless—the product of an unimpregnated egg, becomes the father of countless thousands of worker bees, and of many full-developed queens. The queen with which he mated can, at will, lay eggs of either sex. Passing across the comb from cell to cell she will deposit in one an egg from which will hatch a female (worker), and in an adjoining cell, built larger to accommodate a drone, she will lay an egg that shall produce a male; the former impregnated as it passes the spermatheca **(43)**, the latter, not. Strange, also, that from the egg which the queen, by movement of a muscle has impregnated with element of the male, the workers can, at will, hatch out an undeveloped female like themselves, or a full-developed queen to carry on the reproduction of the species **(197)**. And strange, that eggs laid by a queen who never has been mated, or by a worker who sometimes will rashly take upon her the functions of a queen **(200)**, will hatch out drones, and fecundation follow upon parthenogenesis. **(44)**.

23. A Splendid Example.—The queen, now in "full use," rapidly occupies the cells with eggs, of which from 2,000 to 3,000 may be deposited in one day **(4)**. The population rises. The bees, encouraged by increasing quantities of brood, and urged on by the hunger-wants of growing larvæ, search the country side and carry in rich stores of nectar; still looking to the future; labouring for others; setting a splendid example of diligence, and perseverance, and foresight. Summer will not last for ever. They know it—these patterns of hopeful industry, whose message to the world is wise—"Improve the shining hour, for time in its passing waiteth for none."

CHAPTER IV.

THE BEE IN AUTUMN AND WINTER.

"Morn on the mountain, like a summer bird,
Lifts up her purple wing; and in the vales
The gentle wind, a sweet and passionate wooer,
Kisses the blushing leaf, and stirs up life
Within the solemn woods of ash deep crimsoned,
And silver beech, and maple yellow leaved,
Where Autumn, like a faint old man, sits down
By the wayside a weary."

24. The Death of the Drones.—As autumn with its chill nights and shortening days advances, the supply of nectar rapidly diminishes in the plants. It is an anxious time for the bees. Stores are not accumulating. The colony has suffered serious losses. From time to time the white-sealed combs of honey—the fruit of many days of earnest labour, have been removed, stolen by some dexterous hand. And daily in the combs to which the queen is wedded fresh mouths cry out for food. It is necessary for the survival of the colony that a limit be set to the consumption of stores. The drones—always heavy feeders, and for whom nature has now no sphere of usefulness, have become, by reason of their appetite, the most immediate danger. They have had their day of indulgence, and sunny idleness. Their continued presence in the hive: their death within its portals when the cold of winter should make their removal impossible and render their decaying bodies a source of peril—must be prevented. The time has come for them to share that sacrifice to the future which is the lot of all alike in the community of high ideals to which they belong. In this is no special injustice. Nor can one say, with any degree of certainty, that in this laying down of life for the sake of others there is none of that glorious spirit of love which has inspired the workers to give themselves and all their energies and endurance even unto death, in faithful adherence to their purpose. The slower intelligence of the drone may not realise at once the need that has arisen; and the life of pampered idleness to which, in the nature of things, he has been condemned, may unfit him for that display of voluntary self-abnegation so visible in the other sex. Many, however, leave the hive at noon, never to return. Others, "infirm of purpose," seek to share, for one night more, the comforts of the hive; but sentries at the entrance forbid it and drive them off. Others still, fearful of destiny, have clung to the combs, with weak love of life exceptional in such a race and feeble efforts to resist expulsion to the inhospitable fields without. On them the workers pour the vials of their wrath, and the helpless victims, left by nature defenceless among a multitude of pitiless enemies, succumb to their wounds, or are driven out to join their comrades in misfortune. As the sun sinks and twilight gathers round the scene, the chill of the autumn evening settles upon the vanquished, and all that army of males, once so gay and careless, lies motionless and dead.

25. The Approach of Winter.—That awful tragedy over, the workers return to their more peaceful duties. Blackberry, heather, and ivy still offer their sweets, and much remains to be done before sufficient stores can be collected and sealed to supply the colony with food for winter and early spring. But foraging becomes a more precarious task. The days in which a bee may work out of doors grow shorter. Rain and high winds claim their victims. The strength of the stock diminishes rapidly. And the queen gradually ceases to lay, well knowing that presently the task of the nurse bees will have become impossible. For autumn, with its harvest song and glory tints, is passing, and

> "The leaves are falling, falling,
>   Solemnly and slow;
> 'Caw! caw!' the rooks are calling.
>   It is a sound of woe,
>   A sound of woe!"

Presently silence reigns in the hive. The bees have collected upon the centre combs, clustering closely—

> "Insensibly subdued to settled quiet."

There they will hang together until, the frosts and storms of winter passed, spring shall visit the earth again, and the morning sun, peering through the entrance of their citadel, shall woo them to the work and adventures of another year.

CHAPTER V.

ANATOMY OF THE BEE.

26. A Bee Guide would be far from complete if it failed to supply such information as may appear to be necessary for all who desire to take an intelligent interest in the management of bees. Within the limits of such a work as this, however, it is not possible to enter at any great length into the subject of bee anatomy; nor, indeed, would it be desirable, for, as Hunter has said—

"Of the natural history of the bee more has been conceived than observed. It is commonly not only unnecessary to be minute in our description of parts in natural history, but in general improper. Minutiæ beyond what is essential, tire the mind, and render that which should entertain along with instruction, heavy and disagreeable."—*Phil. Transactions*, 1792.

Those who are anxious to study a subject so interesting, may find all that they require in the various books which deal with it fully.

27. External Skeleton.—The external skeleton of the bee is composed of chitin, covered for the most part with hairs of the same substance, which have their special uses, some as organs of touch, some as brushes, others as gatherers of pollen, or as clothing, protectors, or ornaments. A glance at the illustration (Fig. 7) will show that the body of the bee is made up of three distinct parts, viz.—the head, the thorax, and the abdomen.

28. Head.—The head (Fig. 3) consists of several parts, among which are included the simple eyes; the compound eyes; the antennæ or feelers; and the organs of the mouth.

29. Simple Eyes.—The three simple eyes (*ocelli* or *stemmata*), of which one only is visible in the illustration (Fig. 3), are arranged in triangular form upon the vertex in the queen and worker, and in the front of the face in the drone. They enable the bee to judge accurately of distances out of doors, and to see near objects in the darkness of the hive.

30. Compound Eyes.—The two compound eyes, placed one on each side of the head (Fig. 3), are largest in the drone, and smallest in the worker. They are made up of a number of separate eyes united together, and containing in the drone about 26,000, in the worker, 12,000, and in the queen, 10,000 hexagonal lenses or facets. These, pointing in almost every direction, give to the bee an exceedingly wide range of vision, wider far than would have been possible with a fixed, simple eye.

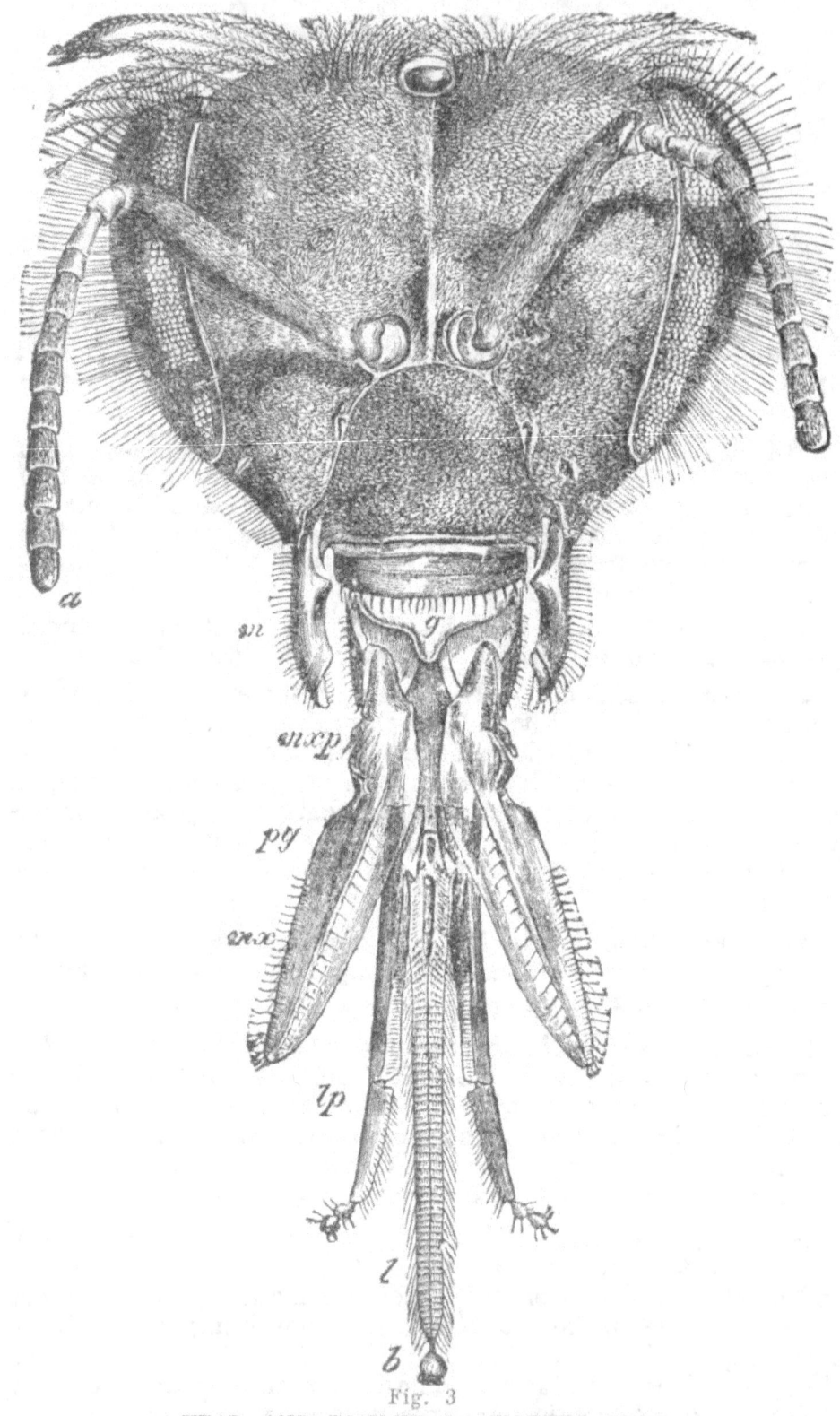

Fig. 3
HEAD AND TONGUE OF WORKER BEE
(Magnified sixteen times.)

a, Antenna, or Feeler; *m*, Mandible, or Outer Jaw; *g*, Epipharynx, or Gum Flap; *mxp*, Maxillary Palpus; *pg*, Paraglossa (shown above the Lingua, opposite *pg*); *mx*, Maxilla, or Inner Jaw; *lp*, Labial Palpus; *l*, Lingua, or Tongue; *b*, Bouton, or Spoon.

31. Antennæ.—The antennæ, or feelers (Fig. 3, a) are cylindrical organs inserted close to each other in the front of the head. They are covered with hairs; and, articulated to the head by a hemispherical joint controlled by three muscles, they can be moved about rapidly in every direction. They are made up of twelve joints each in the worker and the queen, and of thirteen joints in the drone (Fig. 4). The antennæ give to the bee a power akin to that of speech; and, by their motions, form a language in which wants and desires can be communicated.

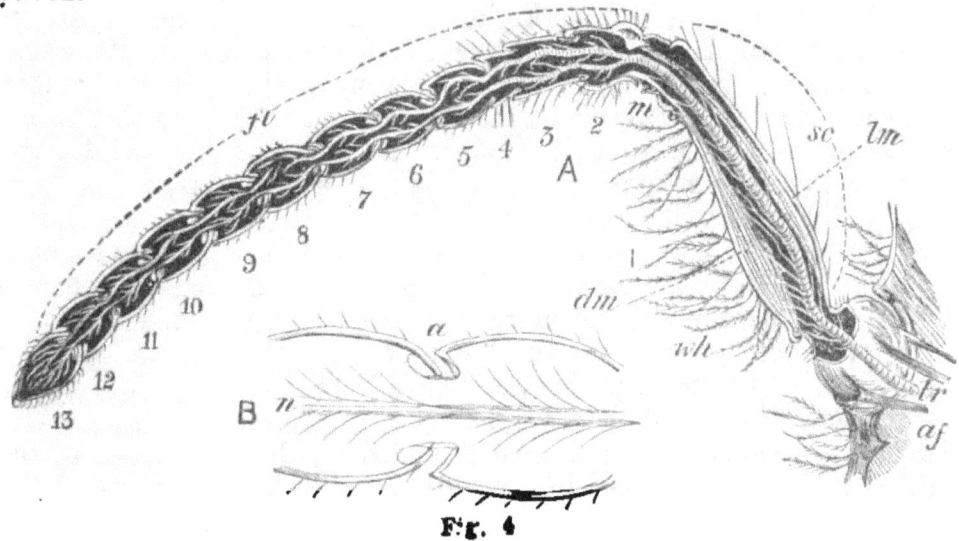

Fig. 4

LONGITUDINAL SECTION OF DRONE ANTENNA.
Nerve Structures removed (magnified twenty times).

A, *sc*, Scape; *fl*, Flagellum; *1, 2, 3, etc.*, No. of Joints; *af*, Antennary Fossa, or Hollow; *tr*, Trachea; *m*, Soft Membrane; *wh*, Webbed Hairs; *lm*, Levator Muscle; *dm*, Depressor Muscle. B, Small portion of Flagellum (magnified sixty times)—*n*, Nerve; *a*, Articulation, or Joint.

32. Organs of Mouth.—The organs of the mouth include the following:—The *mandibles* or jaws (Fig. 3, m) situated one on either side of the labrum. Their movement is lateral. They are provided with hairs, are exceedingly powerful, and, in the queen and drone only, are rough and notched. The *labrum* or upper lip (shown above *q*, Fig. 3), moves vertically. The *epipharynx*, or gum flap (*g*) has a covering of white membrane exceedingly delicate, and is brought into use when liquids are being taken up by the tongue, as explained below. The *maxillæ*, or second jaws (*mx*) are hollowed out, are supplied with very stiff hairs, and, in conjunction with the labial palpi, form a tube in which the tongue works; they bear a short pair of *maxillary palpi*, or feelers (*mxp*). A third pair of jaws— *second maxillæ*, are fused together so as to form a *labium*, or under lip, beneath the opening of the mouth, consisting of a basal *mentum*, paired *paraglossæ* (shown opposite *pg*), by which liquids reach the front of the tongue for swallowing; and *labial palpi* (*lp*) each consisting of four joints, the two terminal joints being very small and supplied with sensitive hairs. These palpi embrace the tongue behind, as the maxillæ embrace it before, and together form a tube surrounding the tongue, as stated. The *lingua*, or tongue (*l*) is connected at its roots with the mentum, and is stretched out or withdrawn by the action of the *protractor linguæ* and *retractor linguæ* muscles. Covering it is a sheath clothed with hairs some of which are sensitive. At the extremity of the tongue is the spoon (*b*), which is provided with delicate hairs. When large quantities of liquid are to be taken up, the tongue, sweeping backwards and forwards by means of a highly elastic rod running through its centre, gathers the liquid upon its hairs; the maxillæ and the labial palpi form a tube around it; and, the front of the epipharynx being lowered to close the space above the maxillæ, the tube is completed to the œsophagus or gullet **(38)**, and the liquid is taken up. When very small quantities of liquid are being taken, the delicate hairs of the spoon, which are capable of gathering up the most minute quantities, collect the liquid and transfer it to grooves at the back of the spoon, from which it is taken up to the paraglossæ, where it reaches the front of the tongue and is swallowed **(58)**. The tongues of the queen and drone are shorter than that of the worker, the last, only, of the three having laid upon her the duty of gathering nectar from the flowers.

33. Thorax.—The thorax (Fig. 7) consists of the three segments below the head, and styled the *pro-thorax*, next the head, and bearing the front pair of legs **(34)**, the *meso-thorax*, in which are articulated the second pair of legs and the first pair of wings **(35)**, and the *meta-thorax*, which carries the third pair of legs and the second pair of wings, and has the first segment of the hind body, or abdomen, **(37)**, fused with it. The thorax is covered with hairs, long and feathered in the worker for the collection of pollen, and in the drone short and spiny, with great power of clinging, but unsuited to the gathering of pollen. The queen is comparatively bare, her mission being confined, chiefly, to the hive.

"A little device will make the bees our assistants in studying their thoracic and leg structure. Take a thin string, about a foot long, and at each end fix a dead bee, by tying round the neck. Drop the suspended 'culprits' between the frames of a stock, so that the middle of the string rests like a saddle on the top bar. In a couple of days, every hair will be cleaned from the 'gibbets,' and their bodies polished like those of beetles, so that the attachment of the wings, the spiracles, the lines dividing pro-, meso-, and meta-thorax, the actual form of the leg joints, and the character of their articulations, with many other interesting points, will be clearly visible."—*Cheshire*.

34. Legs.— Three pairs of legs originate in the thorax—the anterior legs in the pro-thorax; the intermediate legs in the meso-thorax; the posterior legs in the meta-thorax. The anterior leg has the curry-comb—a semi-circular toothed recess, and a *velum*, or sail, by which the antennæ are combed, the legs being moved to the front of the head, and then drawn outwards, cleaning the antennæ which have dropped into the recesses. The intermediate leg is furnished with a spur which has been supposed to act as a lever to remove the pollen balls from the corbicula, but the precise use of which is still a subject of controversy. The posterior, or hind leg of the worker (Fig. 5), consisting of nine joints, is provided, as to the upper joints, with stiff, bristling hairs, by which pollen and propolis are collected. The *tibia* (*ti*) and the *planta* (*p*) are articulated at the inner angles of the joints, and, as they move, the parts opposite *wp* open and shut like jaws, the upper one having a supply of teeth which close upon the lower, flattened surface. These jaws are used for removing the plates of wax from the abdomen (62). They are absent in the queen and drone, wax production being a function of the worker only. The stiff combs (*p*) remove and collect from the hairs of the thorax the particles of pollen

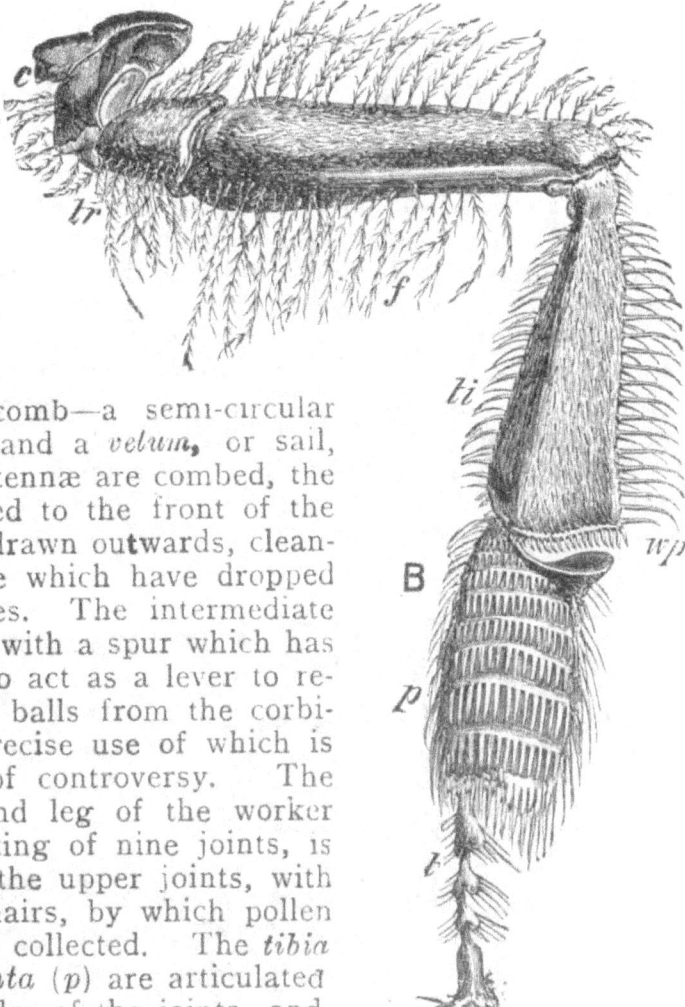

Fig. 5.

THIRD RIGHT LEG OF WORKER.

Side next the body.

(*Magnified ten times.*)

c, Coxa; *tr*, Trochanter; *ti*, Tibia; *wp*, Wax Pincers; *p*, Planta; *t*, Tarsus.

gathered there, and these are transferred to the hollow, fringed portion of the tibia (*ti*) called the *corbicula*, or pollen basket, the combs on the left leg supplying the right corbicula, and those on the right acting similarly towards the left basket. These baskets, with their loads of varied coloured pollen, are familiar objects to all who have watched bees alighting at their hives in the breeding season **(74)**. Corbiculæ do not appear on the posterior legs of either the queen or the drone, the duty of collecting and carrying pollen being assigned to the worker only. The queen, a great walker, has the largest legs, and the drone has the smallest. The *tarsus*, or foot (*t*) has five joints, the terminal joint being furnished with two *unguiculi*, or claws, of great strength, which can be turned up or down as required. These claws enable the bees to cling to their combs, to fix themselves securely to other substances, and also to suspend themselves to the hive-top, or to each other in festoons **(10)** or clusters. Between the claws is the *pulvillus*, or cushion, which secretes an oily, sticky substance that enables the bee to move about upon, or to adhere to glass and other smooth surfaces.

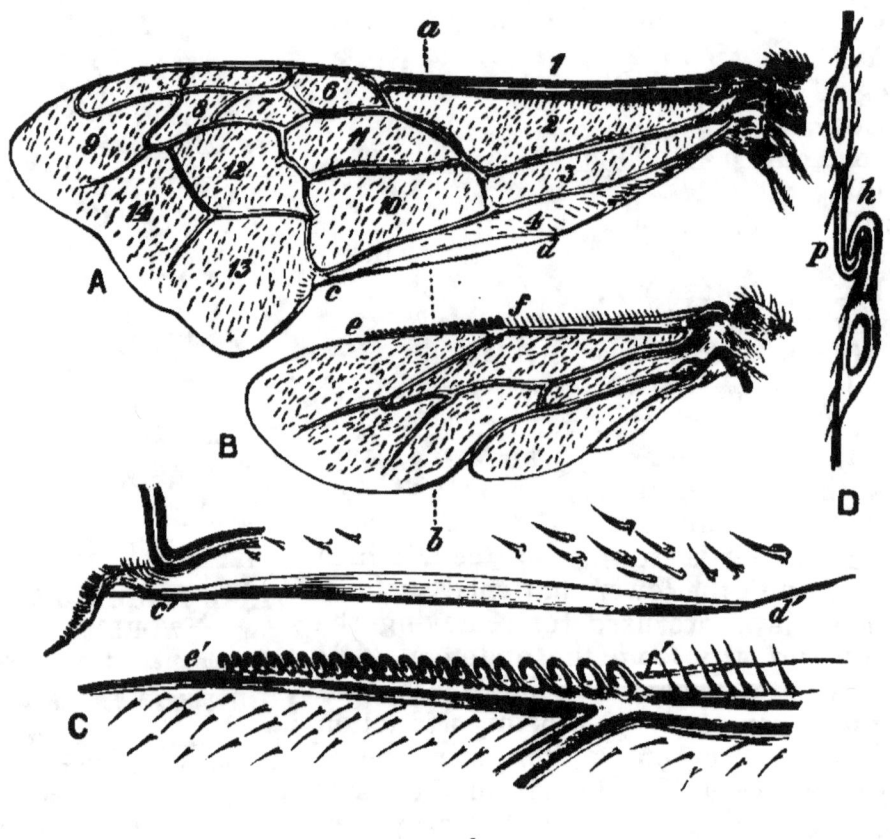

Fig. 6.

WINGS OF THE BEE.—NERVURES, CELLS, AND DETAILS

A and B, Anterior and Posterior Right Wings of Worker (under side), magnified eight times,—*1* to *14*, Cells; *c,d*, Plait; *e,f*, Hooklets. C Plait and Hooklets, magnified twenty-five times—*c',d'*, Plait; *e',f'*, Hooklets. D, Cross Section (through line *a*, *b*,) of *p*, Plait, and *h*, Hooklet, locked together.

ANATOMY OF THE BEE.

35. Wings.—The wings (Fig. 6), which also originate in the thorax, are four in number,—the anterior pair and the posterior pair, articulated into the meso-thorax and the meta-thorax respectively. The upper and outward margin of the posterior wing has a number of hooklets (B. $e, f,$) and the lower and inner margin of the anterior wing is folded in a plait (A. $c, d,$). As the anterior wing is raised for flight, its folded plait passes over the hooklets of the posterior wing and is caught by them (C, and D. p, h), so that the two wings act together as one wing, thus, on the principle—"Unity is strength," adding power and speed to the flight. When the bee alights, the wings become free, and lie closely over the abdomen, thus permitting the insect to enter comb cells, which, otherwise, would be impracticable. The wings of the drone are the largest, and those of the workers the shortest. The vibrations, when in flight, have been calculated by Marey at 190 per second, and by Landois at 440. Bees can fly backwards, and. even when in full flight, can stop very suddenly. When leaving the hive to collect food they will fly at the rate of from fifteen to twenty miles an hour; but, when returning heavily laden, their speed is much less, varying from five to twelve miles an hour. The limit of their usual flight from the hive on foraging duty may be taken as two miles. They have, however, been known, in exceptional circumstances, to travel as far as seven miles in search of food.

36. Spiracles and Tracheæ.—The breathing of the bee is carried on through the *spiracles*, or openings, in the sides of the body (Fig. 7. s) which can be opened or closed at will. These spiracles admit air to the tracheæ, or tubes, which. as shown in the illustration, ramifying in countless number throughout the body, convey the necessary oxygen to the various organs. The development of the tracheæ into *vesicles*, or air sacs, of which the main ones lie in the anterior portion of the abdomen in the worker and drone, greatly assists the bee's flight. When inflated, the air sacs increase the size of the body, thus altering its specific gravity and reducing the amount of effort necessary to accomplish a long and rapid flight **(21).** A bee, at rest, suddenly disturbed, may often be observed to jump, or fly a couple of inches before taking wing, the air sacs not being filled; then, with the lifting of the wings, and rapid extension and contraction of the abdomen, air is drawn through the spiracles into the vesicles; the spiracles are closed, and the insect rises in flight. Immersed in liquid, breathing through the spiracles is stopped, and the insect dies.

22 THE PRACTICAL BEE GUIDE.

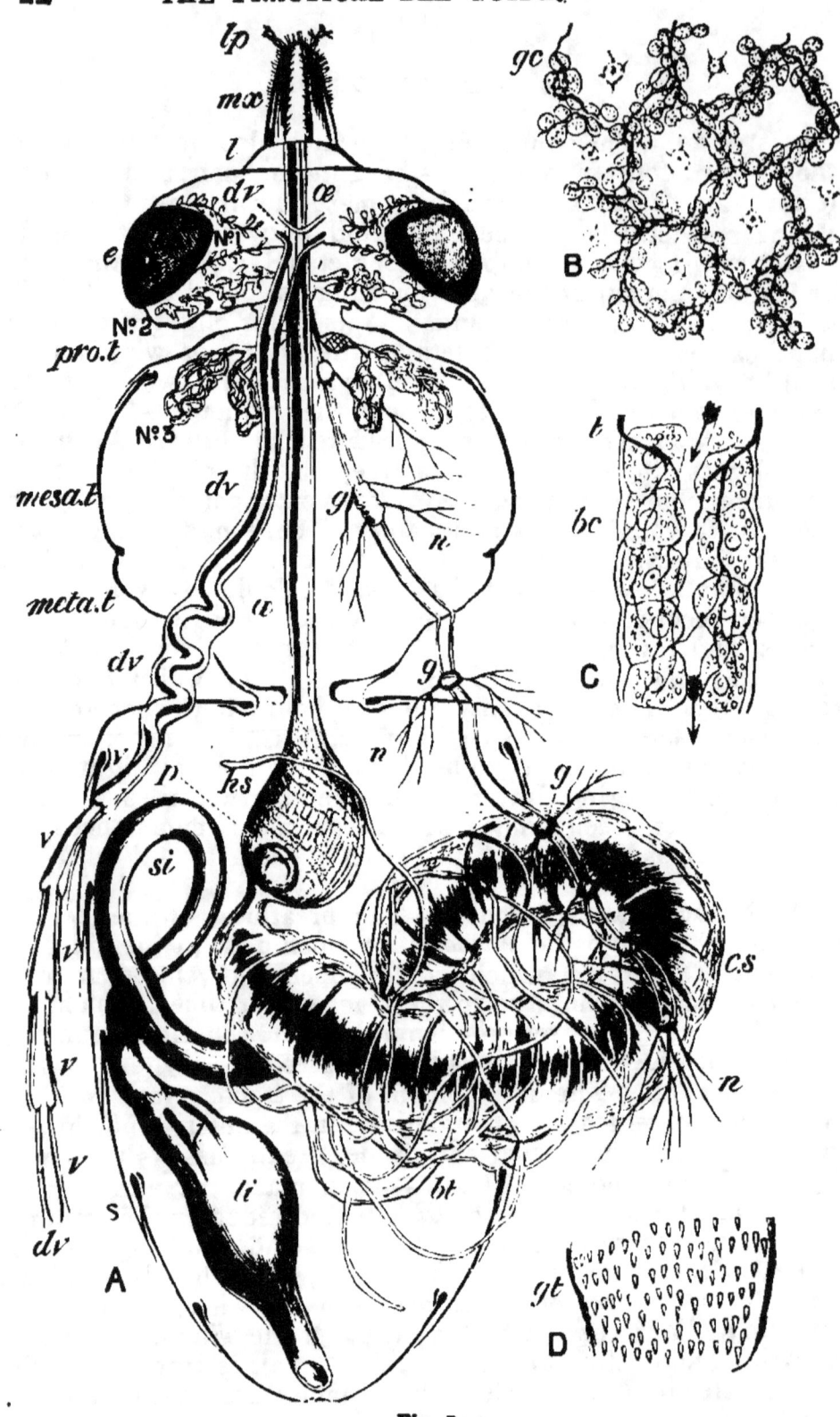

Fig. 7.

DIGESTIVE SYSTEM OF BEE (Magnified ten times).

A, Horizontal Section of Body—*lp*, Labial Palpus; *mx*, Maxilla; *e*, Eye; *dv*, Dorsal Vessel; *v*, Ventricles; Nos. 1, 2, 3, Salivary Gland System; *œ* Œsophagus; *pro.t*, Prothorax; *mesa.t*, Mesathorax; *meta.t*, Metathorax; *g.g.* Ganglia of Chief Nerve Chain; *n*, Nerves; *hs*, Honey Sac; *p*, Stomach-mouth; *cs*, Chyle Stomach; *bt*, Biliary or Malpighian Vessels; *si*, Small Intestine; *l*, Lamellæ; *li*, Large Intestine. B, Cellular Layer of Stomach—*gc*, Gastric Cells (magnified 200 times). C, Biliary Tube—*bc*, Bile Cells; *t*, Trachea. D.

ANATOMY OF THE BEE.

Fig. 8.
UNDER SIDE OF WORKER BEE, SHOWING WAX SCALES.
(Magnified three times.)

37. Abdomen.—Joined to the thorax by a short tube (the *petiole*), is the abdomen (Fig. 7). The worker's is enclosed by six visible rings, or segments of chitin, each of which is constructed of two plates—the *dorsal* plates on the back, and the *ventral* plates on the lower side. Those shown (Fig. 8) are the ventral plates **(10)** where the wax scales are found **(62)**. The abdomen of the queen is longer and more pointed than that of either the worker or the drone, but only in the worker are the secreting membranes present on which wax is produced. (Fig. 1).

38. Honey Sac.—The honey sac (Fig. 7, *hs*) is situated in the abdomen, and is connected above with the *œsophagus*, or gullet (*œ*), running through the thorax to the mouth **(58)**, and below, with the *chyle stomach* (*cs*), beneath which are the *ileum*, or small intestine (*si*) and the large intestine (*li*) or *colon*. Between the honey sac and the chyle stomach is the stomach mouth (*p*) by which, at the will of the bee, the contents of the honey sac may be admitted to, or excluded from the chyle stomach **(58)**. The nectar carried in the honey sac may, by contraction of the muscles there, be transferred as honey through the œsophagus and mouth to the comb cells, or may be admitted through the stomach mouth to the chyle stomach for digestion. The honey sac can hold one-third of an ordinary drop; but the usual load of a foraging bee is only one-fifth.

39. Sting.—The sting (Fig. 9, A) consists of a horny sheath (*sh*) terminating in a sharp toothed edge, and guiding the lancets, or darts (*d, d'*). The lancets have barbed edges (*b, b,*) and are connected above, at *c, c'*, with the compound levers (*i, k, l,* and *i', k', l',*) by which the sting may be forced into comparatively tough substances. When the bee is about to sting, the muscles of the compound levers contracting revolve the latter round the points *f, f'*, and, pressing upwards against the curved arms of

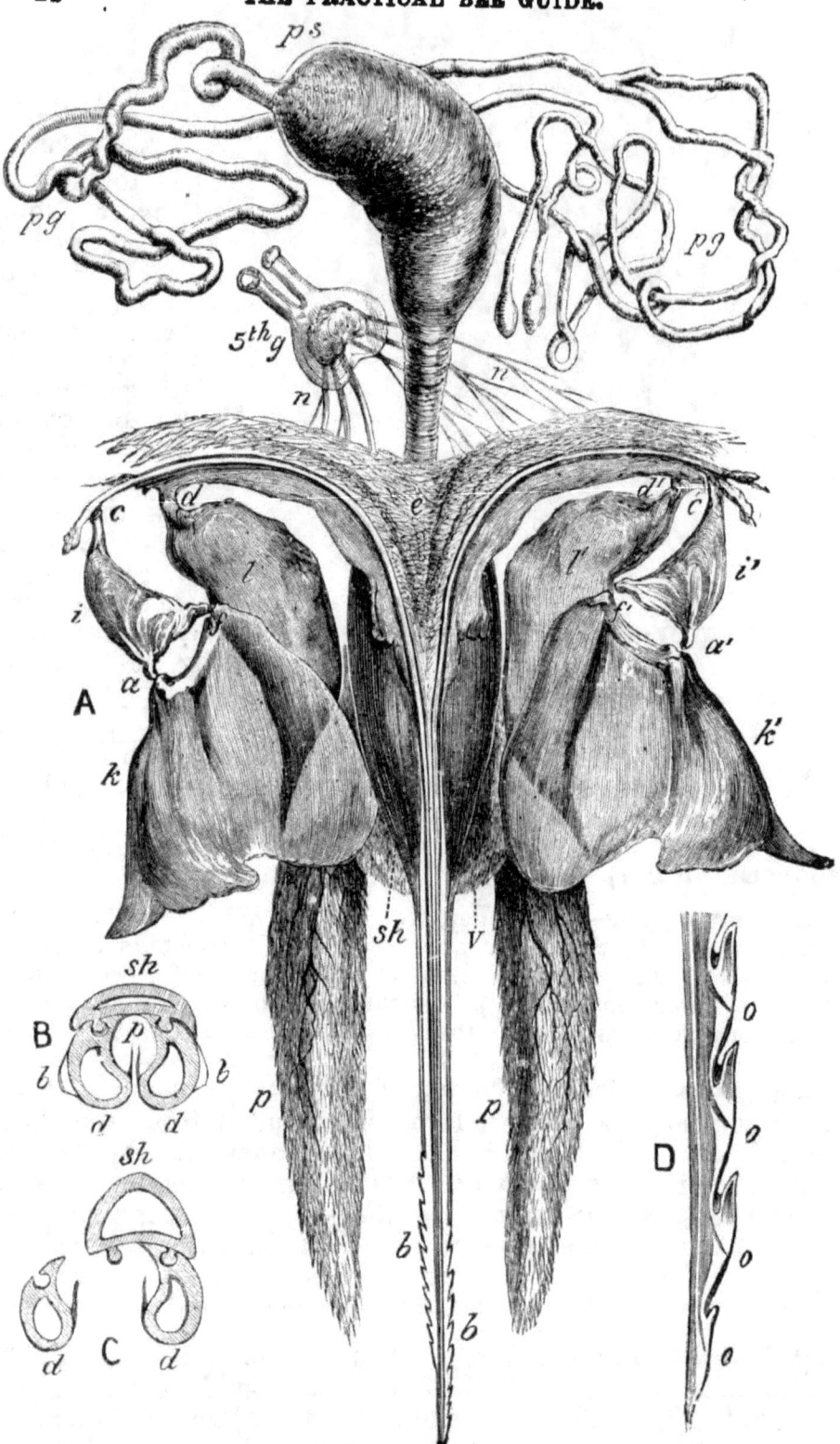

Fig. 9.—STING OF THE BEE.—(Magnified thirty times.)
A, Sting separated from its Muscles—*ps*, Poison Sac; *pg*, Poison Gland, 5th *g*, Fifth Abdominal Ganglion; *n, n*, Nerves; *e*, External Thin Membrane joining Sting to last Abdominal Segment; *i, k,* and *l,* and *i', k',* and *l',* Levers to move Darts; *sh*, Sheath; *v*, Vulva; *p*, Sting Palpus, or Feeler; *b*, Barbs. B and C, Sections through Darts and Sheath, magnified 500 times—*sh*, Sheath; *d*, Darts; *b*, Barbs; *p*, Poison Channel. D, Termination of Dart, magnified 200 times—*o, o*, Openings for Poison to escape into Wound

the lancets at c, c', the levers drive them down with the sheath, and beyond it deeper into the wound. The poison sac (ps) which is supplied by the poison glands (pg) discharges its contents into the sheath, from which the poison is driven, with much force, through channels in the lancets and apertures between their barbs (D. o, o_1) to the lowest part of the wound, until (if the sting be not removed at once) the poison sac has emptied itself. The barbs of the lancets fasten them into the object stung; and, although if left undisturbed the bee is able, by working it round after the manner of a screw, to withdraw the sting (as one might withdraw a gimlet), the pain caused by the injection of the poison generally prompts an immediate assault upon the offender, who, in her effort to escape, frequently leaves her sting in the flesh, and, attached to it, the poison sac and gland. These latter have a reflex action, and, if not removed, may continue to inject poison into the wound for some time after they have been separated from the bee.

40. Palpi, or Feelers.—Quick as is the bee in her attack, she will not proceed to sting until she has examined the surface of the object to be pierced. For this she is provided with the *palpi*, or feelers (Fig. 9, A, p, p), which have sensitive hairs and delicate nerve points, enabling the insect to discover whether the particular spot selected for assault is capable of being pierced.

41. Queen's Sting.—The sting of the queen is longer than that of the worker, and is curved. As already stated **(21)**, the mother bee appears to realise the exceeding value of her life to the colony, and to be unwilling to risk the loss of her sting by incautious use. Unless in very exceptional circumstances, it is not used by her as a weapon of offence or defence, and then only, or chiefly, as against rival queens or other bees **(20)**. The drone has no sting, the sting being an essentially female structure—in reality, a highly modified ovipositor, or egg-laying apparatus.

42. Organs of Drone.—The organs of the drone include two *testes* (Fig. 10. A, t) in communication, by means of the two tubes—*vasa deferentia* (vd) with two seminal vesicles (vs). These vesicles discharge into two *mucus glands* (mg) from which extends the *ductus ejaculatorius* (de) at the end of which is found the organ of generation (o). The *spermatozoa* (B) originate in the testes. As they mature, they pass into the *vesiculæ seminales* (vs) and, mingled with mucus from the glands (mg), proceed continually through the ductus ejaculatorius (de) into the *bean* (b) and, in mass, are called the *spermatophore*. Coition takes place on the wing **(21)** when the pressure of air in the tracheæ and air vessels (h) assists the abdominal muscles in extruding the organs. These, by reason of certain curved rings or ridges (A, r, and E, r') beneath the bean, may not be withdrawn during coition; and, with the expulsion of

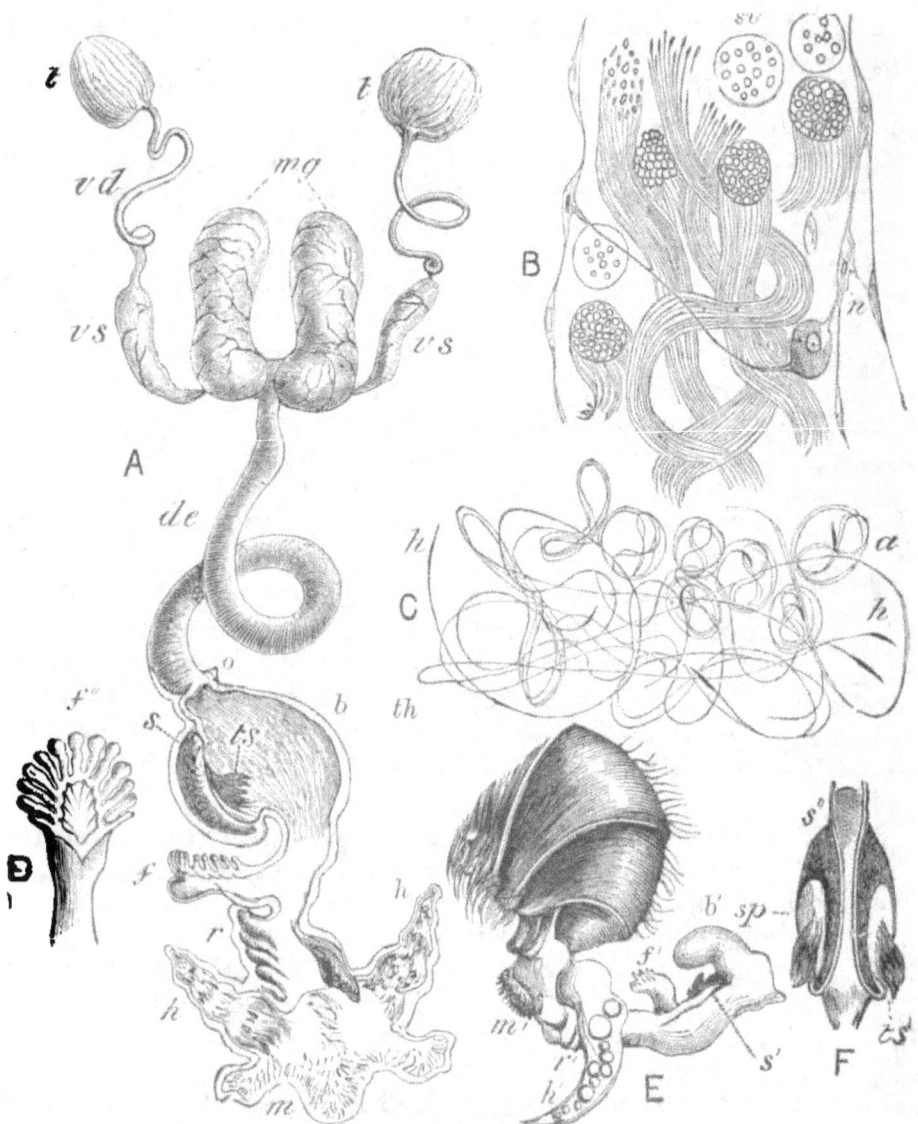

Fig. 16
ORGANS OF DRONE.
(Magnified twelve times.)

A, Organs Removed from Body, but in true Relative Position—*t*, Testes. *vd*, Vas Deferens; *vs*, Vesicula Seminalis; *mg*, Mucus Glands; *de*, Ductus Ejaculatorius; *o*, Termination of Organ; *s*, Sickle-shaped Scale, beneath which Spermatophore is formed; *ts*, Triangular Scale; *b*, Bean; *f*, Fan-shaped Appendage; *r*, Ridges; *h*, Horns; *m*, Masque of Réaumur, or Hairy Membrane. B, Spermatozoa developing within Spermatic Tubes of Testes (Magnified 500 times) —*sv*, Spermatic Vesicle; *n*, Nerve Cells. C, Spermatozoa as they arrange themselves after removal from the body—*a*, Coiled form; *h*, Head; *th*, Thread. D, Face View of Appendage *f* in A—*f"*, Fan-like Fringe. E, Organs Extruded; lettering as A. F, Front View of portion of Bean—*s"* Sickle-shaped Scale; *sp*, Spermatophore; *ts"*, Triangular Scale.

the spermatophore (43) the organs are ruptured, and the drone dies. The queen, having now within her "the potency of the two sexes," returns to the hive carrying, as an appendage, part of the male organs—a sure sign of impregnation.

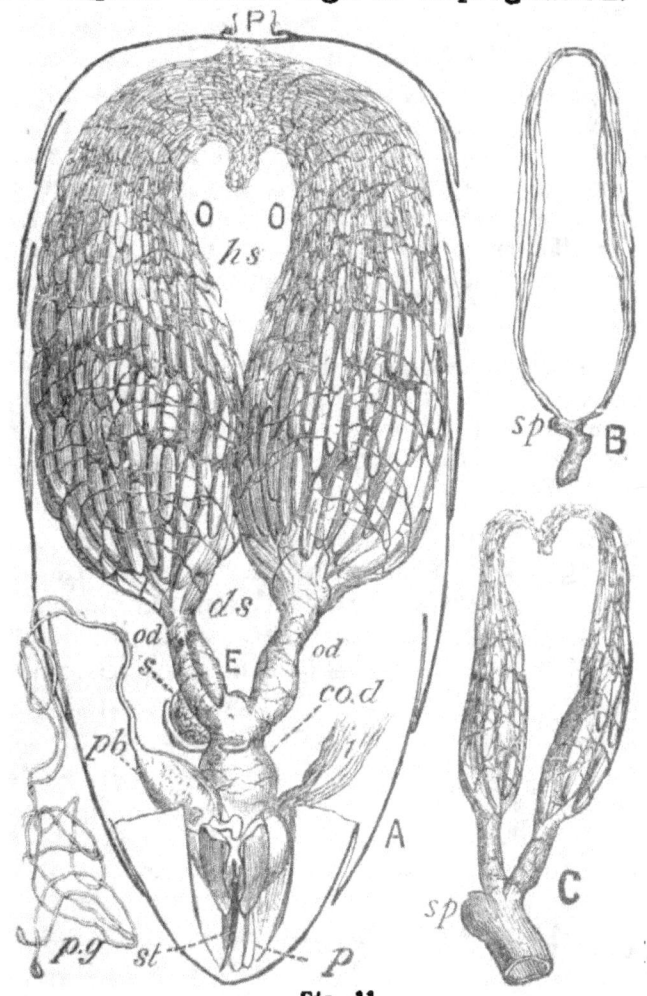

Fig. 11.
ORGANS OF THE QUEEN, ETC.

A, Abdomen of Queen, under side (magnified eight times)—P, Petiole; O, O, Ovaries; hs, Position filled by Honey Sac; ds, Position through which Digestive System passes; od, Oviduct; co.d, Common Oviduct; E, Egg passing Oviduct; s, Spermatheca; i, Intestine; pb, Poison Bag; pg, Poison Gland; st, Sting; p, Palpi. B, Rudimentary Ovaries of Ordinary Worker—sp, Rudimentary Spermatheca. C, Partially developed Ovaries of Laying Worker—sp, Rudimentary Spermatheca.

43. Organs of the Queen.—The organs of the queen include the *ovaries* (Fig. 11, O. O.) in which the eggs are developed; the *oviducts* (od), the *spermatheca* (s, and Fig. 12), which retains the spermatozoa received from the drone and numbering, according to Leuckart, the enormous quantity of

25,000,000 **(22)**; a duct which joins the spermatheca with the vagina, and which, by opening or closing, permits or prevents the passing of the spermatozoa when eggs are traversing the *common oviduct* (*co.d*); and the *vagina*.

44. Parthenogenesis.—In 1845 Dzierzon **(80)** announced his discovery of parthenogenesis in bees **(22)**. In 1849 he wrote—

"In the copulation of the queen, the ovary is not impregnated, but this vesicle or seminal receptical (Fig. 12) is penetrated or filled by the male semen. By this, much, nay all of what was enigmatical is solved,—especially how the queen can lay fertile eggs in the early spring, when there are no males in the hive. The supply of semen received during copulation is sufficient for her whole life. To lay drone-eggs, according to my experience, requires no fecundation at all."

Later on he wrote:—

"All eggs which come to maturity in the two ovaries of a queen-bee are only of one and the same kind, which, when they are laid without coming in contact with the male semen, become developed into male Bees, but, on the contrary, when they are fertilized by male semen, produce female Bees."

45. Fertilisation of the Egg.—The queen can, at will, fertilise the egg as it passes the entrance to the spermatheca, or can allow it to pass unfertilised: in the former case it will produce a female bee; in the latter, a male. It follows that if a queen be mated with a drone of a different race the workers produced by the queen will exhibit characteristics of both parents, while the drones will partake of the nature of the queen only. Exceptions to this rule may, indeed, occur, but very infrequently—as where the drones of a black queen that has mated with a Ligurian drone have shown some slight Ligurian characteristics. Dzierzon and others suggested that these might result from a laying Italian worker, or from the action of an *aura seminalis;* but Siebold proved the existence of seminal filaments in thirty of fifty-two female eggs examined, while in twenty-seven drone eggs similarly examined he found not one seminal filament. The supply of spermatozoa, decreasing as the fertilization of her eggs proceeds, fails and becomes ex-

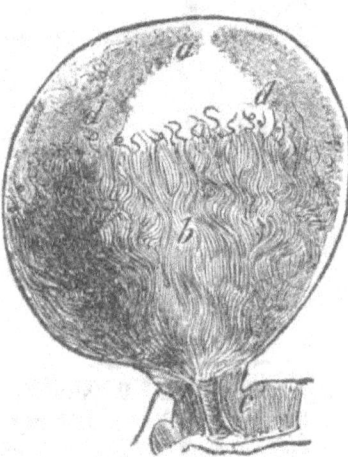

Fig. 12.
SPERMATHECA.
(Magnified forty times.)

a, Space filled by clear fluid; *b*, Mass of Spermatozoa; *c*, Spermathecal Duct; *d. d*, Spermatozoa in activity.

hausted usually at the end of three years; but, even at the close of her second year a queen, under the "forcing" methods of modern bee-keeping, ceases to be profitable, in consequence of the enormous drain upon her resources. **(281).**

"A prolific queen will lay, during her life, 1,500,000 eggs—a number so vast—that the eggs, lying in contact, end to end, would stretch about one and three-quarter miles. A good queen is able to furnish to the cells an average of two eggs per minute for weeks in succession. Taking the lowest estimate, she then yields the incredible quantity of twice her own weight daily, or, more accurately, four times, since at this period more than half her weight consists of eggs."—*Cheshire.*

A queen that has not been mated within twenty-one days of her birth usually becomes incapable of impregnation and a drone-breeder **(188).**

CHAPTER VI.

DIFFERENT RACES OF BEES.

46. Black, or Native Bees are so well known that no description of their appearance is necessary here. They are not so prolific as are some of the other races; but they are hardy, and adapted to our climate. They begin working and breeding early in the spring. They are excellent comb builders, their cappings being white and specially attractive in sections when placed on the market side by side with sections from foreign bees. They cannot always be relied upon to be docile, and easily handled. But they have none of the wickedness of the Syrian **(50)**; they are not inveterate swarmers, like the Carniolan **(48)**; and they are superior to the Ligurian **(47)** as cappers of honey.

47. Italians, or Ligurian Bees belong to North Italy, but are used and valued by bee-keepers everywhere, and in America are exceedingly popular. They differ little or nothing from Black bees in size; but they are lighter in colour, and have three handsome yellow bands beginning with the first segment of the abdomen, by which they may be easily recognised. They are much more prolific than Blacks; are early and late workers; and can collect the sweets from flowers upon which black bees cannot work. They are, however, indifferent comb builders; are often slow to take to supers; and are very capable robbers. They are gentle and easy to manage when pure. But bees from an Italian queen and a Black drone have not the desirable characteristic of amiability, and are generally troublesome in the handling.

48. Carniolans are natives of Austria. They differ in appearance from Black bees, having broad white bands on the lower portions of the segments of the abdomen. They use propolis **(75)** most sparingly, and build beautifully white combs. They winter well; begin work early in spring; and, although very gentle, are stout defenders of their homes. The most amiable of bees, manipulation of their hives can be carried on with ease and confidence, and on this account they are very suitable bees for beginners **(180)**. The objection to them is, that, owing to the exceedingly prolific nature of their queens, they are inveterate swarmers when kept in small hives, and exposed to the sun in hot summers. This, however, is an objection which will not weigh with those who desire rapid increase, and who are capable of exercising due control over the swarming propensity. **(216).**

49. Cyprians have been introduced into this country from Cyprus. Their bodies are smaller and more pointed than those of Black bees, and the three yellow bands are continued under the abdomen, as they are not in the Ligurian race. They are extremely prolific, and diligent workers; but their comb is too inferior to justify their use for the production of section honey. Laying workers **(200)** are more frequent among them than among Blacks. Their lavish use of propolis adds much to the difficulties of managing them; and they are so vindictive that they have been adopted only in very exceptional cases in this country.

50. Syrians differ little in appearance from the former. For queen-rearing purposes they are valuable, because a queenless colony of Syrians will build a large number of queen cells—sometimes as many as thirty on one frame **(196).** It follows that if eggs or unsealed larvæ of any race be given to a queenless colony of Syrians, the production of queens may be enormously increased. They are wicked, most difficult to handle, and are often quite unmanageable.

51. Giant Bees (*Apis dorsata*) are found in India, Ceylon, China and Eastwards to Java. They build single combs, five or six feet long by three or four feet deep, in high trees or rocks, remaining only two or three weeks in one place, and travelling sometimes 100 miles to make a new home. They are exceedingly wicked, often inflicting fatal injuries upon man and beast, and offering little encouragement to any attempts at domesticating them.

52. Common East Indian Bees (*Apis Indica*) are common in India and from Madagascar to the Malayan Archipelago. They are small, yellow underneath the abdomen, and not difficult to manage. Their production of honey, under the methods by which, to a limited extent, they are worked in their native country, does not often exceed fifteen or twenty pounds.

53. Dwarf East Indian Bees (*Apis florea*) are the smallest honey bees known. They are black, with the anterior part of the abdomen a bright orange. Their combs seldom exceed eight inches in length by four inches in depth, and the cells are so diminutive that 100 are contained in a superficial square inch of comb. Their production of honey is too small to render their cultivation profitable.

54. Dutch Bees vary in colour from brown to black, not infrequently showing three tan-colour bands on the first three segments of the abdomen—due, apparently to an introduction of Italian blood—the remaining segments being brown, or black. Dutch beekeepers encourage swarming, and their bees respond, refusing to adapt themselves at once to our methods. Twenty standard frames may be allowed for the brood nest; or ten, or twelve, with a crate of shallow frames on top. But Dutch bees may be expected to swarm, in spite of all precautions, if the queen be more than six months old. They are usually as good tempered as Blacks. Their combs are regular, with white cappings, attractive in sections. They were introduced to these countries in the hope of counteracting the ravages of I. W. disease **(360)**.

55. Sand Bees (*Andrena*) are found in this country, occasionally in large numbers. They differ from the honey bee in many structural points, notably in their much shorter tongue. The females are always fully developed, so that the "worker" caste does not exist among them. They make their nests by burrowing in the ground, usually in sandy places. Although they are "solitary" insects, in the sense of forming no social communities like those of the honey bees, a large number of nests are generally found close together, and many individuals may be seen, in the spring months, flying around their favourite haunts.

56. Leafcutter Bees (*Megachile*) are long-tongued, like the honey bee, but they may be distinguished by their broad head, powerful mandibles, and generally stout build. Like the *Andrenæ*, they have no "worker" caste. They nest in the ground, sometimes digging burrows, but more frequently using ready-made hollows, such as the tunnels of worms. They neatly cut pieces out of the leaves of plants, and use these to build their nests, in which they store food for their grubs. The nest resembles a number of thimbles placed inside one another. These bees also nest in old trees and walls. They are rarely found in the North. Their nests have been discovered among the quilts of bar frame hives; and Mr. M. H. Read has found them twice in his apiary, and frequently in the keyhole of his garden-door.

56b. Caucasians are natives of the Caucasus, in Russia. Within the past few years they have been recommended in the United States by the Department of Agriculture, and testimony to their exceeding gentleness and prolificness has been given by many prominent beekeepers. Neither smoke, carbolic, nor protection is necessary when these bees are being manipulated; they show little resentment when roughly treated; their queens are great layers, and their workers are exceptionally industrious.

CHAPTER VII.

BEE PRODUCTS, &c.

57. Honey.—It is a common error to suppose that honey is gathered by bees from flowers. Honey is the product of the nectar secreted in the nectaries of flowers, and subjected to a chemical change in the honey sac **(38)** of the bee; the cane sugar of the nectar being converted into the grape sugar of honey by its mixture with the secretion of certain glands in the insect. Speaking generally, nectar may be said to contain from 50 per cent. to 80 per cent. of water **(59)**, according to the flowers from which it is collected and to the state of the atmosphere as damp or dry. Some flowers—the fuschia, for example, secrete nectar which has a much smaller percentage of water. The secretion is nature's provision for securing the fertilization of plants by inducing the visits of insects, notably of the bee, in order that pollen, the fertilising dust, may be carried from flower to flower **(74)**. It is affected by temperature, and by the state of the weather. It is lessened by continued drought, and increased by gentle rain accompanied by heat. Usually it is greatest in the morning; decreasing in the afternoon. Every bee-keeper knows what it is to have his bees idle during days of sunshine, tho' situated in the midst of honey-producing plants and flowers, when long absence of rain and dew has retarded the secretion of nectar.

58. Gathering and Storing Honey.—When bees visit the flowers, they suck the nectar by means of the spoon **(32)** and groove; and, passing through the œsophagus or gullet, it enters the honey sac **(38)**. Below the honey sac is situated the stomach mouth which the insect can, at will, open to admit the honey to the chyle stomach as food, or close when the honey is intended to be stored **(38)**. In the latter case the muscles of the honey sac are brought into play, and the fluid is forced out of the mouth and deposited in the comb cells. The existence of the honey sac and stomach mouth explain various phenomena in the life of the bee—how, when swarming, she can carry from the

(Photo by J. G. Digges.)
Fig. 13.
BEE ON CLOVER.

hive sufficient honey to serve as food for a considerable time, and even for the production of wax in her new home (18); how, in the winter season, she can feed from the contents of the honey sac during several days without having recourse to the comb-cells.

59. Water in Honey.—When nectar, thus converted into honey, has been deposited in the cells, it becomes necessary to evaporate from it a quantity of water. Dr. Smyth says:—

"In order to complete a pound of sealed honey in their comb-cells the bees must evaporate at least half a pound, and frequently a pound of water from the cells, and out of the hive."—*Irish Bee Journal.*

This is done by raising the temperature of the water and of the interior of the hive, and by fanning (14) the moist and heated air out through the doors. On occasions of extreme humidity of the atmosphere outside, evaporation within the hive becomes arrested and the gathering of nectar ceases for a time. When a cell is almost filled with honey, it is sealed with a capping of wax, and in that condition the honey will keep indefinitely in a warm, dry place.

60. Honey as Food.—As an article of food, honey is very valuable. It requires no digestion; is a great heat producer; a gentle laxative; and a purifier of the blood.

61. Honey Dew.—An unpleasant, dark, rank-flavoured substance called honey dew, is sometimes gathered by bees, much to the annoyance of their owner. During a spell of hot, dry weather, with absence of moisture and rain, this objectionable deposit may be seen upon the trees, and the bees eagerly gather it. Its name is due to an erroneous opinion by which it was described as a dew of honey falling upon the leaves. Investigation, however, has shown that the substance is a discharge from the bodies of aphides, which suck the sap of certain trees, and discharge it continuously as a saccharine, viscous fluid. In the absence of rain to wash it off, it adheres to the leaves, and is resorted to by both bees and ants. Ants (368) are particularly fond of it, and may often be seen literally milking the aphides. It is stated by Lubbock, who made a special study of the subject, that certain species of ants "farm" aphides in their nests, feeding them with the leaves required, and enjoying the saccharine produce of these "milch cows." Honey dew is sometimes produced without any action of aphides, as an exudation (*Miellée*) from the leaves.

62. Beeswax.—Beeswax has a specific gravity of between .960 and .970, and will melt at 144° to 148° Fahr. It is a natural secretion, produced in a liquid state by the wax glands in the body of the bee, and moulded, in the shape of tiny scales, in the wax pockets under the ventral plates (**37.** Fig. 8). From these pockets the scales are transferred to the mouth, to be made flexible previous to being used in comb building **(10).** The wax scales are "so thin and light that one hundred of them hardly weigh as much as a kernel of wheat."--(*Dubini*). For the secretion of wax, bees require a temperature of from 90° to 95° Fahr. They feed liberally, and then form in clusters, remaining inactive in a high temperature until, after about twenty-four hours, the honey, converted into wax, appears as described above.

63. Honey used in Wax Production.—Just what quantity of honey is required by clustering bees for the production of wax, it is not possible, with our present knowledge, to state definitely. Opinions upon this subject vary considerably. Until further discoveries have been made, it may be taken as a fairly accurate estimate, that, according to the conditions existing in the hive, from 10 lbs. to 16 lbs. of honey are consumed by clusters which produce 1 lb. of wax. If honey be valued at 6d. per lb., and wax at 1s. 8d. per lb., it follows that from 5s. to 8s. worth of honey is used in the manufacture of 1s. 8d. worth of wax, to which must be added the severe strain upon the bees which wax production imposes, and the cost of the devotion to that work of so many bees who might be more profitably occupied elsewhere **(73).**

64. Paraffin Wax and Ceresin Wax are mineral products, unsuitable for bee hives. They are sometimes used for the adulteration of beeswax by manufacturers of foundation, but, being of a lower specific gravity than that of beeswax, their presence as adulterants may be easily detected **(114).**

65. Honey Comb.—The combs of a hive at swarming time will be found, on examination, to contain four distinct kinds of cells, viz.—Worker cells; Drone cells; Transition cells; and Queen cells **(183).**

66. Worker Cells (Fig. 14, O), in which worker bees are reared, are about $\frac{1}{2}''$ deep, and $\frac{1}{5}''$ wide; so that five cells measure about 1", and from twenty-seven to twenty-nine go to the square inch **(187).**

67. Drone Cells (Fig. 14, F), in which drone bees are reared, are about $\frac{3}{4}''$ deep, and $\frac{1}{4}''$ wide; so that four cells measure about 1", and from sixteen to eighteen go to the square inch **(194).**

68. Hexagonal Cells.—Both worker and drone cells are six-sided, or hexagonal—a shape which gives the greatest capacity and strength with the least expenditure of material and labour

(10). If they were built square or triangular they would not be so adapted to the shape of the bees to be reared in them: if they were circular, much valuable space would be sacrificed between the touching circles: being hexagonal they approximate to the shape of the bees; avoid waste of space; and so support each other that they can be constructed of the lightest material, and of exceeding delicacy, for

"Walls so thin, with sister walls combined;
Weak in themselves, a sure dependence find."
—EVANS.

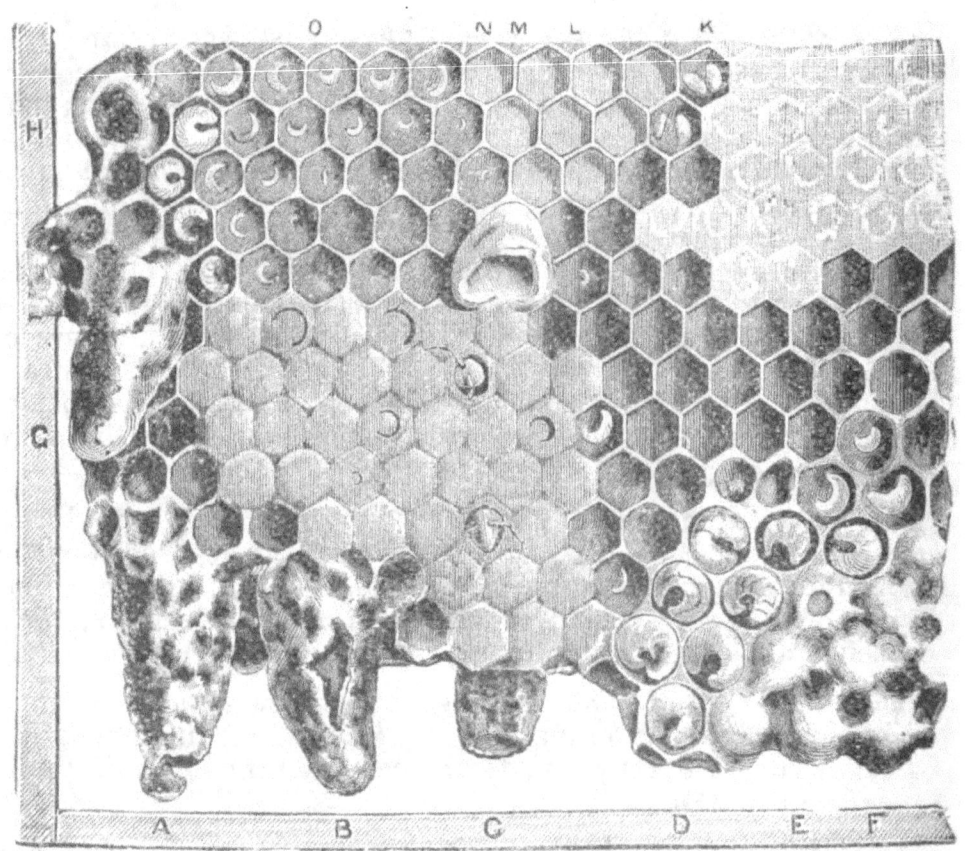

Fig. 14.—HONEYCOMB.—(Natural size)

A, Queen Cell, from which Queen has hatched, showing lid; B, Queen Cell, torn open; C, Queen Cell, cut down; D, Drone Grub; E, Drone Cell, partly sealed; F, Drone Cells, sealed; G, "False" Queen Cell, and beyond—Worker Cells, sealed, and Bees emerging from cells; H, Old Queen Cell; I, Sealed Honey; K, Fresh Pollen Masses; L, Cells nearly filled with Pollen; M, Aborted Queen Cell on face of Comb; O, Eggs and Larvæ in various conditions.

69. Transitional Cells, irregular in shape,—better called Intermediate or Accommodation Cells—are constructed to connect worker and drone, or queen, cells.

70. Use of Cells for Storing.—The three kinds of cells—Worker, Drone and Transitional, may be used for storing honey and pollen. They slope upwards from the base, thus being easier to fill, and safer as receptacles for honey than if built horizontally.

71. Queen Cells (Fig. 14, A, B, C) are built much stronger than the cells already described. They are made of a mixture of wax and pollen, the pollen being introduced to render them porous **(196)**. They are like waxen thimbles, about an inch long, and tapering downwards **(17)**. Unlike ordinary brood cells, queen cells are not used a second time, but are cut down by the bees (Fig. 14, C, H) usually within a few hours of the birth of the queen.

72. Cappings.—Cells occupied by brood have a porous capping of wax and pollen; and those which contain honey are capped with wax.

73. Value of Combs.—The wax employed in the combs of 11 ordinary "standard" frames **(97)** weighs about 2 lbs. According to the estimate made elsewhere **(63)**, 2 lbs. of wax represent the consumption of from 10s. to 16s. worth of honey: and if to this be added the value of the time occupied by the bees in secreting the necessary wax, and in building the combs, the strain upon their constitutions, and the loss of honey which, in the season, they might have gathered if not occupied otherwise, the value of the combs to the bee-keeper may be estimated at from £1 to £1 10s., perhaps considerably higher. Comb is, therefore, a thing too costly to be wasted; and the more use the bee-keeper can take out of his combs, and the more economically he can have them built, the more profitable will his industry be **(113)**. It must, however, be stated that combs should not be used indefinitely for breeding purposes, because the portions of cocoons left in the cells by hatching bees **(191)** eventually reduce the size of the cells so appreciably that they become no longer suitable for brood rearing **(190)**.

74. Pollen.—Pollen is the fertilizing dust of flowers, and for bees, an indispensable food. On examination of a typical flower (Fig. 15, B) it is found to be composed of four whorls, or sets of organs on the same plane with one another and distributed in a circle about an axis. These organs are:—(1) The outer whorl, or *calyx* (a): (2) the second whorl, or *corolla* (b): (3) a whorl of parts alternating with the corolla, and called the *andrœcium* (c): and (4) the inner whorl, or *gynœcium* (d). Nos. 1 and 2 are the floral envelopes or coverings. No. 3—the *andrœcium*—is made up of a series of leaves, or stamens (A): these are the male organs, and have at their summits the anthers (*e*) which contain the fertilizing dust (*f*). No. 4—the *gynœcium*, or *pistil* (C)—is the female sexual organ, situated in the centre of the flower, and containing the ovary (*g*) and stigma (*h*). For the production of a perfect seed it is necessary that the germs of the pistil be fecundated by the pollen of the stamen. When the pollen grains are ripe they are shed by the anthers. Some flowers are bisexual, or hermaphrodite, having both male and female organs, but these are rarely self-fertilized, nature having provided against it: as, for example, in the primula, in which occur long stamens and a pistil with a short style, or short stamens and a pistil with a long style. Fertilization is commonly effected by insects, and to encourage their visits, the perfume, nectar, and gay colours

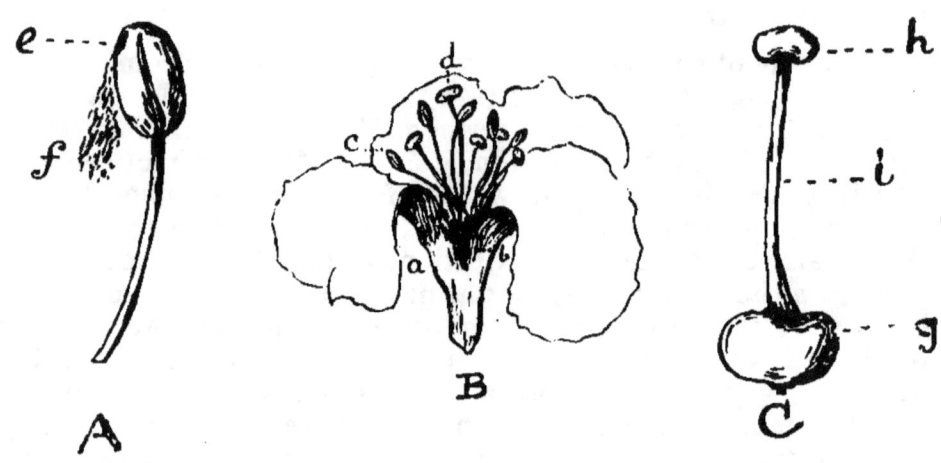

Fig. 15.
FLOWER: STAMEN: AND PISTIL.

A, Stamen, *e*, Anther; *f*, Pollen. B, Sketch of Typical Flower, *a*, Calyx:—Sepals; *b*, Corolla:—Petals; *c*, Androecium:—Stamens; *d*, Gynoecium (Pistil):—Carpels. C, Pistil, *g*, Ovary; *h*, Stigma; *i*, Style.

of the flowers are developed **(57)**. Bees are among the most useful workers in this field of nature's economy. They enter the flowers **(9)**, gather their loads of pollen, even roll themselves in the rich dust, fill with it their baskets (*corbiculæ*) **(34)** carry home their provender and deposit it in the cells. Thus the pollen is carried from stamen to pistil, and from flower to flower, and fruits and flowers become fertilized which, but for the visits of the insects, would remain barren. Bees have long been recognised as valuable fellow-labourers with the horticulturist; and many cases are on record in which fruit trees have ceased to bear, or have borne but indifferently, when bees had disappeared from the neighbourhood. When pollen arrives at the hive, it is packed in cells, is often covered with honey, and is sealed over with wax.

"Having gone to visit the bees of a lady friend living within six miles of Dublin, we first went to look at her fine peach house. She pointed out to me how badly the blooms had set on a tree that was not easy to fertilize, and said that the gardener had been complaining that the bees had not helped him as much as usual this spring by visiting and fertilizing the blooms. A look at the hive soon explained this, all the combs, except the outside ones, being a compact mass of rotten foul brood, a very few live bees being left in one hive."—M. H. READ in *Irish Bee Journal*.

75. Propolis.—Propolis is a resinous, sticky substance, gathered from pine, horse chestnut, and other trees, and carried by the bees, as they carry pollen, on their hind legs. It is used for filling up cracks, to exclude draughts, and to make the hive watertight. It is applied also, much to the bee-keeper's regret, to fasten together the frames, and other parts of the hive furniture **(174).** When hives are situated under high trees, the vibration, caused by the roots as

"Through woods and mountain passes
The winds, like anthems roll,"

is felt at once by the bees, who endeavour to modify it as far as possible, by fixing their combs and frames in the hive with propolis **(9).** Occasionally bees will use propolis to defeat their natural enemies, or to fix and render harmless unwelcome intruders. The term, "Propolis," signifies "before the city"; the use of the substance in the defence of the hive having been observed. Huber describes the construction of barricades of wax and propolis in the hive entrance, to exclude the Death's Head Moth (*Sphinx Atropos*), while giving passage to the workers. Reaumur observes that a snail having gained admission to one of his hives, the bees, being unable to remove it, promptly arrested its progress by fastening it down with propolis. Maraldi relates a somewhat similar occurrence, his bees having covered all over with propolis a large slug which they had been unable to dislodge. It is quite a common practice with bees in modern hives to attach pieces of Naphthaline **(352)** to the floorboard, and even to enclose them in a case of propolis to overcome the objectionable smell.

75b. Adulteration of Honey.—When honey has been adulterated with glucose, the presence of glucose may be recognized if a little of the mixture be slowly poured into a glass containing absolute alcohol, and if the alcohol then shows turbid, or milky, having a gummy substance at the bottom.

PART II.

HIVES AND APPLIANCES.

CHAPTER VIII.

HIVES AND FRAMES.

76. Ancient Hives. Bee-keeping as an industry is ancient. It is certain that from a very early period a high value has been set upon honey as an article of food, and that, long centuries ago, bees were kept for profit in manufactured hives. Virgil, who wrote B.C. 70, describes the hive in use in his day. It was constructed of plaited osiers and bark, and was plastered with mud to make it waterproof. Pliny tells us that when the spring flowers in the Italian valleys had failed, the bees, in their hives, were carried at night up the rivers in boats, in search of better pasturage. In parts of Asia hives of pottery were used, and were built into the walls of the houses. The osier hive of Virgil was, probably, somewhat like the old-fashioned straw skep, with which we in this country are so familiar.

Fig 16.
THE SKEP.

HIVES AND FRAMES.

77. The Skep (Fig. 16) is made of straw, sometimes worked on a frame of hazel-rods or cane. It has its uses: but as a permanent home for bees its defects are too many and serious to admit of its adoption by anyone who desires to keep bees for profit, and upon humane principles. It does not permit proper management. It does not allow that perfect control of the bees and their work which is essential to success. Although its cost may not exceed a couple of shillings, it is expensive, because it precludes the use of foundation **(110)** which, in the modern hive, effects so large an economy. It is dangerous, because, not open to examination by the owner, it may harbour disease without his knowledge, and may spread infection far and near **(327)**. And, associated as it is with the hateful sulphur pit, by which our forefathers, for want of a better method, obtained the honey harvest by sacrifice of the bees who gathered it **(142)**, it is not to be encouraged as an adjunct of modern bee-keeping, except within certain limitations.

Fig. 17.
THE SKEP, FLAT TOP FOR SUPERING.

78. Uses of Skeps.—Bees crowded in skeps are likely to give off early swarms, and, with that object in view, stocks in skeps may often be turned to good account. Skeps are useful also for carrying swarms **(153)**, and in the operations of driving **(160)** and hiving **(234)**. They may be used, to some extent, for the production of surplus honey in supers **(271)**. For this purpose the skep is made with a flat top (Fig. 17) having a hole in the centre of the crown.

79. The Skep giving place to the Moveable-Comb Hive.—But modern bee-keeping encourages more intelligent management, and aims at higher success than can be hoped for by the exclusive use of the straw skep—now, happily, giving place to the hive with moveable frames **(81)**, which has effected a revolution in bee-keeping by admitting adequate supervision over the work of the colony; by facilitating the harvesting of larger quantities of honey; and by rendering unnecessary, indeed inexcusable, the destruction of the bees.

80. Genesis of the Moveable-Comb Hive.—Towards the close of the eighteenth century, Huber **(142)**, the blind Naturalist, who was born in Geneva in 1750, constructed a hive in shape

Fig. 18.
THE "C. D. B." HIVE.

like a book, each leaf containing a comb, and by this means he was able to arrive at the discoveries which have made his name famous. In 1838, Dzierzon, a German and the discoverer of parthenogenesis, *i.e.*, reproduction without fecundation **(44)**, began the use of hives in which the combs were attached to top bars. This was improved upon in 1851 by Langstroth, "the father of American apiculture," who invented the hive opening at the top, and with combs in moveable, suspended frames.

81. Advantages of the Moveable-Comb Hive.—To this invention modern bee-keeping owes the rapid progress it has made in the past half century. The moveable frame gives free access to all parts of the hive, and admits of the various operations by which control is exercised over the bees, and their labour turned to the best account **(142)**. The condition of the colony may be thoroughly inspected **(327)**; bees and combs may be changed from one hive to another as required **(252)**; queen rearing **(286)** and artificial swarming **(222)** may be practised; natural swarming controlled **(216)**; honey extracted without destruction of the combs **(276)**, and such intelligent management can be pursued as may produce the best results.

82. The Hive in General Use in Ireland.—The modern moveable-comb hive in general use in Ireland (Figs. 18 and 19) is made to take the frame which has been adopted as the "standard" frame **(97)**. The external measurements of hives may vary to any extent; but the internal measurements of the brood nest, or body box **(86)**, must be such as will with the utmost accuracy suit the measurements of the frame to be

Fig. 19.
THE "FEDERATION" HIVE.

used, and must provide such bee space (83) as careful observation of the natural instincts of the bees has shown to be desirable. This appears to be too obvious to require explanation. Yet some unfortunate mistakes have been made by inexperienced persons in manufacturing hives to a given external measurement, only to find that the frames could not be worked in them.

83. Internal measurements. — The internal measurements of a modern hive are too exact to admit of slipshod carpentery. A 1-16th of an inch, one way or another, may make or mar a hive; and an inaccuracy of a nature so trifling that it would be quite inconsiderable in the case of a piano or of a wooden leg, may render a hive utterly useless for the keeping of bees upon modern principles. A moveable-comb hive is such only when its combs are moveable; and it is found that if the spaces between the ends of the frames (97) and the inner walls of the body box (86) are less than ¼-inch, the bees, being unable to pass, will fasten the frames to the body box with propolis, while if the spaces are more than ⅜-inch, the bees will build brace comb there. There is, therefore, a safe space from ¼-inch to ⅜-inch, and if this be increased or diminished the frames are liable to be fastened to the body box, in which case manipulations of the hive will involve unnecessary exasperation of both the bees and their keeper. As to the respective advantages of the ¼-inch and the ⅜-inch spaces, some difference of opinion exists among experienced bee-keepers. All, however, agree that where bees are found to respect the ⅜-inch space, that space offers very important advantages in the greater facility with which frames may be moved, and the minimising of the risk of crushing bees, and even of killing queens, during manipulations **(182)**. Between the bottom bars of the frames and the floor board **(85)** a space of ⅜-inch should be left. It follows from what has been said that accuracy in the making of hives is essential. If it be desired to manufacture hives at home, one good hive, as a pattern, should be procured, and the measurements of that hive, so far as the internal dimensions are concerned, should be followed with the utmost exactness. The timber should be of good quality and thoroughly seasoned. American seasoned pine is largely used in the manufacture of the best hives. All wooden hives require to be kept well painted to protect the timber from the effects of the weather.

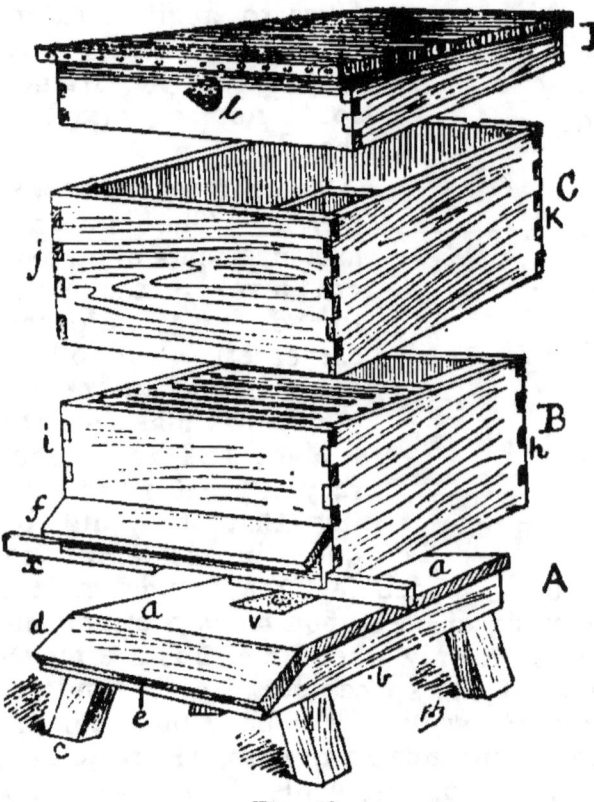

Fig. 20.
THE "FEDERATION" HIVE.
(The Parts Separated.)

84. The "Federation" Hive (Figs. 19 and 20) consists of four parts, viz.:— Floor board and Legs (A); Body box or Brood chamber (B); Lift or "Riser" (C), and Roof (D).

85. The Floor Board (Fig. 20, A) is made of two pieces of timber (*a, a*) 16⅞″ × 11″ × ⅞″, rabbeted ½″ and nailed or screwed to two rails (*b*) 23½″ × 2″ × ⅞″, to which rails the legs (*c*) 8¼″ × 3″ × 2″, are fastened. The rails are chamfered to 20¼″ at the front ends, and on the chamfers is nailed the alighting board (*d*), 16⅞″ × 4½″ × ⅞″; the upper edge chamfered to an angle of 60°; and the lower edge rabbeted ⅜″ × ⅜″ (*e*) to form a rest for a hiving board **(233)**. In the floor board a round hole 2″ in diameter is cut as a ventilator (*v* and Fig. 21), and is covered on the upper side with perforated zinc, the under side having a piece of wood 7″ × 3″ × ⅜″ screwed to the floor board, so that it can be revolved to open or close the ventilator at will **(91. 218. Fig. 111.)**

86. The Body Box (Fig. 20, B) measures, internally, 18" in length, 17" in width and 9½" in depth. This is sufficiently large to take eleven frames and one dummy **(93)**. The sides (h) are 19¾" × 9½" × ⅞", the front (i) is 17¾" × 8⅝" × ⅞", and the back, 17¾" × 9½" × ⅞", rabbeted ½" × ½" at the bottom to rest on and overlap the floor board. These are dovetailed, and are nailed together flush on the upper edges. If put together with a double rabbet nailed, the above measurements must, of course, be altered accordingly.

Fig. 21.
FLOOR-BOARD VENTILATOR.

The front and back have two grooves running from top to bottom, ⅜" wide × ¼" deep, beginning 1⅛" from the ends. Two inner walls, 18½" × 8½" × ⅜", are fitted into the grooves and are chamfered outwards on the upper edges to carry the frames (Figs. 22, 39,): they are nailed ¼" below the top of the body box, so that when the body box is placed in position upon the floor board the sides and back overlap the floor board, and the front, being only 8⅝" deep, leaves a space of ⅜" between it and the floor board as an entrance for the bees. The inner walls being ¼" below the level of the sides and back, and ¾" distant from the sides, the frames **(97)**, when in position, are level with the top of the body box, and are prevented by the hive sides from moving laterally **(267)**. It will be seen that the body box measures internally when complete 18" × 14¾" × 8½". This leaves a space of ⅜" between the frame ends and the inner walls of the body box, and a space of ½" between the bottoms of the frames and the floor board. Four slips, 17¾" × ¾" × ¼", are nailed between the tops and bottoms of the inner walls and the hive sides, and a space of ⅜" is left between the upper slips and the bottoms of the frame shoulders.

Fig. 22.
CHAMFERED INNER WALL.

The spaces between the inner walls and the hive sides are sometimes filled with cork dust, chaff or sawdust, to preserve the heat of the brood chamber. The front (i) has either a ¼" groove in the centre of the bottom, or an arrangement in the porch in which run two doors (x) 8" × ½" × $\frac{3}{16}$" so that the entrance may be reduced, or enlarged, or closed as required. Above the doors a porch (f) is provided to keep off rain from the entrance. (See also **91**, page 48.)

87. The Lift or Riser (Fig. 20, C) measures internally 20" long, 18" wide, and 12" deep, and is made of two pieces 20¾" × 12" × ⅞", and two pieces 18¾" × 12" × ⅞" dovetailed together. Four pieces, 3″ × ⅞″, are screwed to the insides, ¾″ from the bottom. When in position the lift overlaps the body box, and in winter it is reversed and telescoped over the body box, thus providing additional walls, and assisting to preserve the heat of the brood chamber. The lift, in summer, serves to enclose the supers **(99)**, and allows sufficient space for packing round them. When the lift is reversed for winter, the porch is removed from the body box, and is fastened to the lift, in a corresponding position **(378** and Fig. 116, page 207**)**.

88. The Roof (Fig. 20, D) measures internally 19⅞″ long × 18⅛″ wide × 5½″ deep in front, and 4½″ deep at the back. It is made of two pieces, 21⅛″ × 5½″ running to 4½″ × ⅞″ for the sides; one piece, 19⅝″ × 5¼″ × ⅞″, for the front; and one piece, 19⅝″ × 4½″ × ⅞″, for the back. The front and back are rabbeted ¼″ to overlap the lift. On these are nailed two pieces, 23″ × 10¾″ × ⅞″, which are covered with zinc, thus making the roof perfectly rain and snow proof. The roof slopes to the back to throw off rain. It makes a convenient table for the smoker, and other appliances, when neighbouring hives are being manipulated. When in position the roof overlaps the lift. Two holes, 1½″ in diameter, are cut in the gables; the front hole having two escape cones (*l*) fitted to it, to permit the exit of bees which otherwise might be imprisoned, and the back hole being covered inside with perforated zinc. These holes act as ventilators, and the cones are sometimes used for the purpose of clearing bees from supers **(273)**. Roofs are also made A-shape (Fig. 23). In this case the front, back and sides are the same length as the roof described above. Four pieces, 24⅝″ × 6¼″ × ⅞″, feathered to ⅜″, form the cover, overlapping each other 1″; and a ridge board, 24⅝″ × 2″ × 1¼″, cut out ⅝″, is fitted on the top. The objection to such roofs lies in their tendency to open at the joints, and to admit damp. They should be kept well painted. (See **91**.)

Fig. 23.
THE "A" ROOF.

89. The "W.B.C." Hive (Fig. 24).—This hive, which is so popular in England, consists of the following separate parts, as illustrated:—Stand; floorboard; body box (A), to take 10 standard frames and a division board; a 9″ cover (B), with porch and doors, to enclose the body-box, leaving room for packing; super (C), to take 10 shallow frames and a division

HIVES AND FRAMES.

Fig. 24. THE "W.B.C." HIVE.

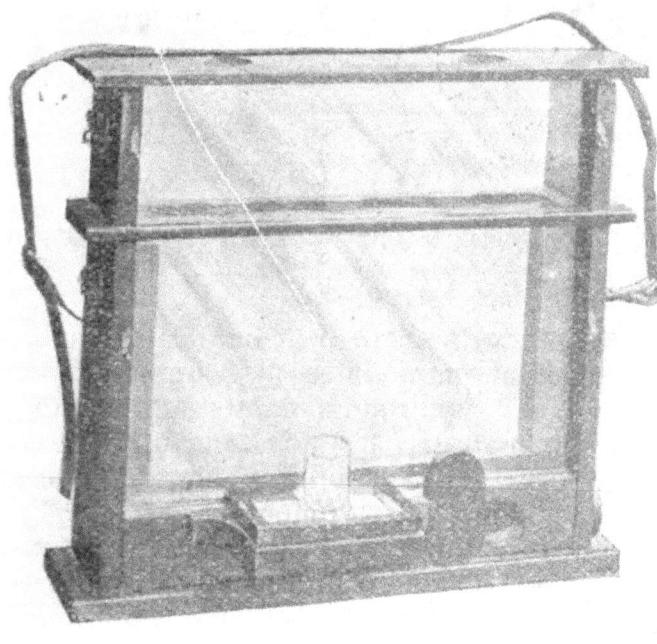

Fig 25. "BRICE" OBSERVATORY HIVE.

board; two 6" covers (D) to enclose supers; a 3" eke (E) to be used under the body-box for wintering, or under a shallow frame super to accommodate standard frames; roof (F) with a 3" lift attached, and fitted with cone escapes. The hive was designed by the late Mr. W. B. Carr.

90. Observatory Hive. — The "Brice" Observatory Hive (Fig. 25) takes one frame below, and four sections or one shallow frame above. The sides are double-glazed, and are fitted with baize-covered shutters. A feeding arrangement is supplied at the side. The floor is round, and can be revolved at will to permit dead bees to drop into a receptacle underneath. There are ventilators with shutters which work on pivots, an exit for the bees, and a strong strap with which to carry the hive. Such observatory hives can be had to hold two "standard" frames and six sections. They are very suitable for use at exhibitions and shows; and are most useful to those who desire to study the hive-labours of the bee.

91. "I. B. A. 1909" Hive.—In 1909, the committee of the Irish Beekeepers' Association adopted specifications for an improved hive, which were published in full in the *Irish Bee Journal* for June and July of that year. The FLOOR BOARD of this hive is made level, so that the hive body may be moved back to admit a feeder on the "Alexander" principle **(123)**; the "J.G.D." ventilator (Illusn. p. 197, and Fig. 111, p. 198) is $4'' \times 8''$, is covered with perforated zinc, and has a graduated sliding door underneath, working from the back of the hive; the rabbet of the alighting board **(85)** is increased to $\frac{3}{4}''$, in order to provide better support for a hiving board. The HIVE BODY has inner walls of $9''$, and outer walls of $10''$ (except the

Fig. 26.   Fig 27.
THE "I.B.A. 1909" HIVE.

front, which is $9\frac{1}{2}''$, so as not to overlap the floor board), and measures $19'' \times 22\frac{1}{2}''$, which accommodates 13 frames and a dummy, while the extra width facilitates manipulation with the lift **(87)** in position, and permits better packing of supers **(268)** to preserve warmth; a slip at each side, to stop the ends of the frames, reduces the inside measurement there to $17''$; a slip, $\frac{3}{8}''$ deep and $\frac{1}{16}''$ thick at bottom bevelled to $\frac{1}{4}''$ thick at top, is attached to the upper inside front, leaving $\frac{1}{8}''$ space between each end of the slip and the opposite edge of the frame shoulder, so that the usually neglected outer edge of the front comb may be worked out by the bees, while the passing up of bees between the front of the lift and the super is prevented. The DOORS (Fig. 26) are supported by brass screws working in $7\frac{1}{2}''$ slots, so that they can neither jam nor drop out. An

ALTERNATIVE ENTRANCE (Fig. 27) has been approved, 12" × ⅜", cut out of the floor board 3¼" from the front, a cut of 1½" × ¼" being taken from the floor board to admit the doors; no porch is required for this entrance; the doors cannot drop out; in case of robbing, the carbolic cloth treatment (310), permitting entrance at one side only, can be more easily applied; having no projecting porch or alighting board, the hive can be more safely conveyed by rail, or otherwise. The DUMMY (93) has its top bar projecting ⅛" inwards, so that the side of the comb next the Dummy may be properly worked out by the bees. The LIFT is 11" deep, made of 11" timber, ¾" thick front and back, and ½" thick for the sides, if nailed, or ½" thick front and back, if dovetailed; the inside slips rest evenly on the four sides of the body-box, and so that the lift, when inverted, does not rest upon the porch, and the way is barred against ants, earwigs and other intruders. The ROOF is flat, 6" in front sloping to 5" at back (including the overlap), is made of timber of the same scantling as the lift, is covered with zinc turned underneath, grooves under the four overlaps preventing water running backwards; the projection is 1½" front and back, and ¾" at the sides; double cones are fitted in the roof front, in 1¼" hole chamfered on the inside to permit the inner cone to fit home to its flange—an important detail when cones are being used as bee escapes (273). The nails used are "non-rusting," and the entire hive is put together with white lead.

92. The "Hibernian" Hive.—This hive, as to the body box, is made on the lines of the "C.D.B." hive, with a porch extending across the entire width of the front. The lift is 11" deep, and telescopes over the body box. The roof is flat, and zinc-covered. The hive takes eleven frames and three section crates, and is dovetailed throughout.

93. The Dummy or Division Board (Fig. 28) is made to fit the body-box (86). It consists of a piece, 14⅝" × 8½" × ½",

Fig. 28.
DIVISION BOARD, OR DUMMY.

Fig. 29.
ENAMEL CLOTH, FOR END OF DUMMY.

with a top bar of 16" × ⅞" × ½", to run on the chamfered tops of the inner walls. It is, therefore, level with the tops of the frames. Two plinths, 8½" × 1⅛" × ⅜", are nailed to the back to prevent warping. The Dummy is ⅛" less than the width of the body-box, and the ends are fitted with two slips of enamel cloth, 1½" wide, to fill the spaces, and to conserve heat (Fig. 29). The enamel cloth may be folded, and fastened between the plinths and the Dummy. The Dummy is less likely to be fastened at the ends by propolis when enamel cloth is used.

94. Use of the Dummy.—Dummies are used for enlarging or contracting the brood nest as required **(236)**. By their means the hive can be adjusted to the size of the colony, and frames can be removed and replaced with greater ease to the manipulator and with greater safety to the bees **(182)**.

Fig. 30.
"FEDERATION" DUMMY.

95. "Federation" Dummy.—Dummies can also be used for ventilating the brood nest during very warm weather — a necessary provision when it is desired to control the swarming propensity **(218)**, and also for feeding, comb-cleaning, and other purposes **(278)**. To supply this want, the "Federation" Dummy (Fig. 30) has been devised. It has a piece, 10½" × 4½", cut from the bottom. The vacancy may be filled with perforated zinc, or excluder zinc **(109)** as required. The plinths are rabbeted, and a slide, 11½" × 5", with ends rabbeted to correspond with grooves in the plinths, slides between the latter, and can be raised, held at any point, and lowered as desired.

"It is most ingenious; and a very valuable addition to bee appliances. How many times I could not tell, that I have met with a comb so bad that I have taken it out of the hive, notwithstanding the fact that there were quite 100 worker brood in it; because, had I left it in till they were hatched the queen would have more eggs deposited in it. But with such a Dummy it would only need to place the faulty frame behind and raise the slide, and remove the frame when the brood was hatched out. For back feeding and ventilation it is also of use. But most of all it is of use when treating a foul broody stock. When it is desired to remove frames containing some diseased cells, the brood frames can be placed behind this dummy, and the queen given two or three frames of foundation in front, and in 21 days the diseased frames can be removed."—TURLOUGH B. O'BRYEN, in the *Irish Bee Journal*. (See Illus. p. 197

96. Sheet and Quilts are required upon the frames or supers to preserve heat; to prevent draught; and to keep the bees from ascending into the roof. The sheet is made of bed ticking or unbleached calico. The quilts should be of felt, carpet, or other warm material. The sheet and quilts should be large enough to cover the interior of the body-box when they are placed upon the frames. From the sheet a circular piece may be all but cut out from the centre, so that it can be turned back when feeding is in progress **(119)** to give the bees access to the feeder; at other times it can be restored to its original position. The sheet, lying as it does upon the frames, should not be made of woollen material, because bees are apt to catch their claws **(34)** in such stuff, and that irritates them. The sheet will lie flat upon the frames if put on damp in the first instance. In summer, a sheet of American cloth, enamelled side down, may with advantage be used instead of a sheet of ticking; but at other seasons it is advisable that the covering should be of porous material to permit evaporation of the moisture of the hive. **Straw** mats or chaff cushions are some-

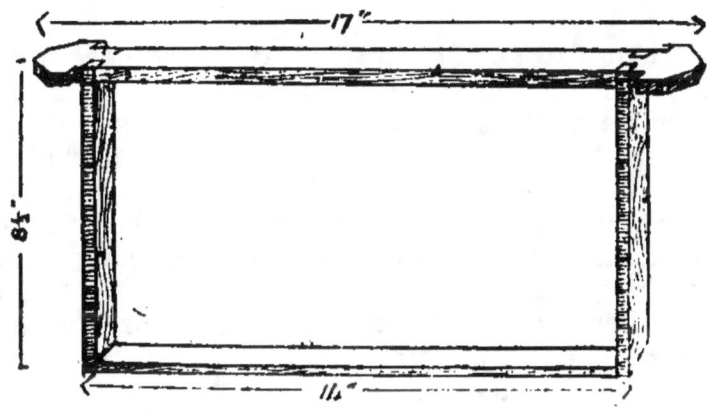

Fig. 31.
THE "STANDARD" FRAME.

times used on the quilts. In winter, it is useful to cover the brood-nest and quilts with an empty crate or other bottomless box, having a piece of canvas or calico tacked underneath, and filled with cork-dust, chaff, or other warm material **(378)**.

97. Frames.—The frame which is here described has been, for some years, recognised in this country as the "standard" frame (Fig. 31). It is made for use in a hive, the measurements of which have been given above, and in accordance with the known instincts of bees, as applied to the building of their combs. We observe that worker comb, i.e., comb in which worker bees are reared, is $\frac{7}{8}$" thick; the frame is therefore made $\frac{7}{8}$" wide. The spaces between sealed brood combs are about $\frac{1}{4}$" to $\frac{3}{8}$"; the frame is therefore intended to provide a $\frac{3}{8}$" space between the combs. This is effected by shoulders on the top bars of the frames, or by the use of "metal ends," by which, when the frames are pressed together in the hive, the necessary space is provided. The frame measures 14″ long × 8½″ deep. The top bar is 17″ long × ½″ thick; the side bars are 8½″ long × ¼″ thick; and the bottom bar is 14″ long × ¼″ thick; the width of all being, as already stated, ⅞″. The four pieces are made to dovetail into one another, and are usually sold in the flat. When put together they should be fastened at the corners with four tacks or fine wire nails. Underneath the top bar are two grooves, the centre groove to hold an edge of a sheet of foundation **(117)**, and the side groove to take a thin wedge supplied with the frame, and by which the foundation is held in the centre groove. Frames are now sold at prices so low that it is not advisable for bee-keepers to manufacture frames for themselves. It is necessary that the frame be put together perfectly square. There are also in use frames having a saw-cut along and through the top bar, into which the sheet of foundation is fastened **(117)**, and frames with plain top bars to which the foundation is attached by melted wax **(117)**.

98. Various Sizes of Frames.—Frames are used of larger size for the brood nest by some bee-keepers, and it is claimed for the larger frames that they give better results. The practice generally in this country is to use the "standard" frame, as described. It is of importance that, whatever size be adopted, it should be uniformly used in the apiary, because there is a decided advantage in being able to interchange frames. In America the popular frame is larger than our standard frame, and many bee-keepers at home hold that our standard should be enlarged. The Langstroth frame, in use in America, has the following dimensions:—Top bar, 19⅛″ long × ⅞″ thick; side bars, 8⅞″ long × 5/16″ thick; bottom bar, 16¾″ long × ⅜″ thick; the width of all the bars being ⅞″. The "Simplicity" frame, which is described as the "Standard frame of America," has the top bar 19⅛″ long; side bars 9⅛″; and bottom bar 17⅝″. Super, or "Shallow" frames, for use in extracting supers **(108)** are in very general use. They differ from the standard frame in being only 5½″ deep, the super being 6″ deep. It is claimed for them that they are more readily taken to by the bees in supers than are standard frames, as they increase the accommodation above the brood nest more gradually; but it is an objection to them that they are not interchangeable with standard frames.

98b. The "Claustral" Detention Chamber.—This appliance, which was illustrated and described in the *Irish Bee Journal* for October and November, 1906, was devised by M. l'Abbé Gouttefangeas, whose book, "Ruche Claustrante et Méthode

CHAPTER IX.

APPLIANCES FOR SUPERING.

99. Supering.—The term "supering" is applied to the use of sections and frames above the brood nest, in order to obtain surplus honey **(255)** of a marketable quality, and free from the mixtures of larval remains and pollen which used to characterise the honey offered for sale before modern bee-keeping introduced better methods.

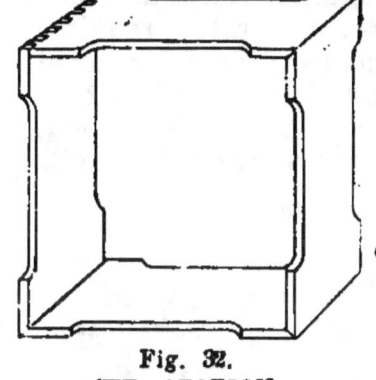

Fig. 32.
THE SECTION.

100. The Section (Fig. 32) is a square case of bass wood, $4\frac{1}{4}'' \times 4\frac{1}{4}'' \times 2'' \times \frac{1}{8}''$. It is sold flat, in one piece, dovetailed at the ends, and with three V-shaped cuts across the wood (Fig. 33) to permit the folding of the section. Bee-ways are provided by reducing the width of the wood to $1\frac{5}{8}''$, so that when the sections are pressed together in the crate **(103)** the bees can pass in and out of them.

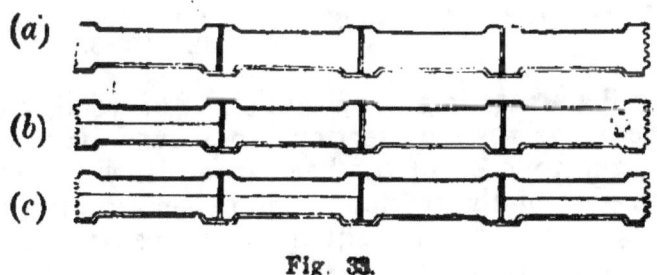

Fig. 33.

(a) FOUR-WAY SECTION; (b) SPLIT-TOP SECTION; (c) THREE SPLIT SECTION.

101. Sections of various kinds are now supplied, viz.—Two bee-way which have spaces provided top and bottom; four bee-way (Figs. 32 and 33, a, b, c) which allow spaces on all the sides; ordinary—not split; split top (b) which have a cut along the centre of the top to grip the foundation **(110)** when it is inserted; three-split (c) with the cut carried through the top and both sides, which secures the foundation on three sides and allows three sections to be fitted with foundation at one operation (258); and four-split, which are supplied in two pieces. The latter are not much used in this country. Sections, of the size indicated above, hold, when filled, one pound of comb honey, and are those for which the public seem to have a preference. It is desirable to have the sections so split that the dovetail comes on the top when the section is in position, for this minimises the risk of opening the section when removing it from its crate. "Tall" sections, measuring $5'' \times 4\frac{1}{4}'' \times 1\frac{3}{8}''$, and holding one pound of comb honey, and sections to hold two pounds, are on the market, but they have not come into general use. The objections advanced against the "tall" one pound section are, that it needs a deeper crate; requires foundation of a special size; is extravagant, requiring more foundation than the smaller section; and that the mid rib of the comb is thicker, in proportion, than in the $4\frac{1}{4}'' \times 4\frac{1}{4}''$ section, and therefore not so likely to be unobserved by the eater (112).

Fig. 34
(a) SHORT SEPARATOR.   (b) LONG SLOTTED SEPARATOR FOR 4 BEE-WAY SECTIONS.

102. The Separator is a very thin sheet of wood, zinc, or tin, used between the rows of sections to secure even surfaces to the combs, and to prevent the bees from drawing out the cells beyond the edge of the sections. Short separators (Fig. 34, a) are $4\frac{1}{4}'' \times 4\frac{1}{4}'' \times \frac{1}{18}''$, i.e., square with the sections. Long separators (Fig. 34, b) are $12\frac{3}{4}'' \times 4\frac{1}{4}'' \times \frac{1}{18}''$, covering three sections. Bee-ways are cut out to permit the bees to pass freely from one section to another. The long separators are easier to handle, and those made of zinc or tin will, with ordinary care, last for many years. Wooden separators, being so thin, require careful handling to avoid breakages.

103. The Section Crate, or Rack (Fig. 35), is a bottomless box constructed to hold twenty-one $4\frac{1}{4}'' \times 4\frac{1}{4}''$ sections, a follower (106), and springs or wedges. It is made of two pieces $17'' \times 4\frac{3}{4}'' \times \frac{5}{8}''$, and two $14\frac{1}{4}'' \times 4\frac{3}{4}'' \times \frac{5}{8}''$ dovetailed together. Its internal

APPLIANCES FOR SUPERING.

measurements are, therefore, 15¾" × 12⅛" × 4⅞". The ⅛" extra in depth is intended to allow for shrinkage, for it is of great importance that the crate, when in use, should not be in the least degree shallower than the sections; otherwise, when crates are tiered up on the hives **(269)**, the weight resting upon the lower sections tends to depress the laths on which they stand, and to destroy the bee space, thus leading to serious mischief **(83)**. Underneath, a frame of ¼" laths is placed.

Fig. 30.
SECTION CRATE.

These carry the sections and separators, and when the crate is placed upon the frames, the laths provide the necessary bee space between the frames and the sections. It follows that if the laths are less than ¼" or more than ⅜" thick, the bees will fasten the sections to the tops of the frames, and serious difficulty will arise when it becomes necessary to remove the crate **(269)**. Crates are sometimes fitted with tin or zinc bars instead of laths, to bear the sections. These, however, are so easily put out of shape, or "dinged," that they can be successful only with very careful handling.

Fig. 33.
DIVISIONAL CRATE.

104. The Divisional Crate (Fig. 36) consists of three single crates, each holding seven sections. It is used towards the close of the honey flow, to secure the perfecting of unfinished sections, when the bees would not have time or honey to fill a larger number **(269)**.

105. Observatory Crates (Fig. 37) are constructed so as to hold sections, separators, and a glass follower, with or without springs. A door is made in the end of the crate, and when it is opened the sections can be seen, and an opinion can be formed as to the state of the work in the crate.

Fig. 37.
OBSERVATORY CRATE.

106. The Follower is a piece of timber $12\frac{3}{4}'' \times 4\frac{3}{8}'' \times \frac{5}{8}''$, and for the Divisional Crate, $4\frac{1}{4}'' \times 4\frac{3}{8}'' \times \frac{5}{8}''$. It is inserted in the crate immediately after the last row of sections to press them together. It is kept in its place by springs or wedges (Figs. 35 and 36).

107. The Hanging Crate, or section frame (Fig. 38) is used for holding six sections in the body box or super box. Separators are attached to the frame on both sides to prevent the

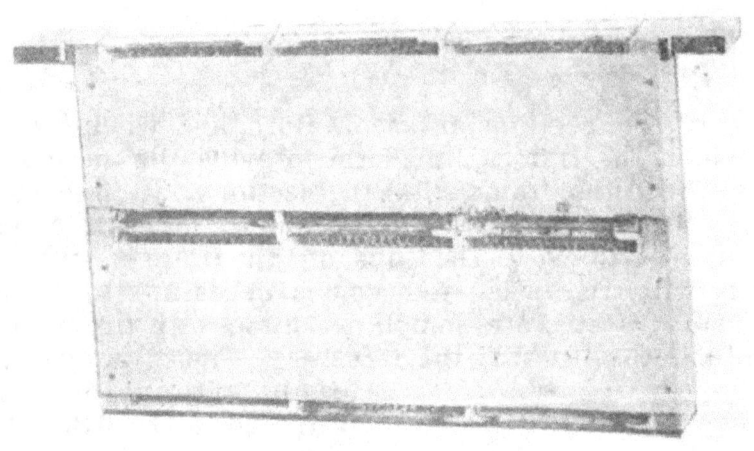

Fig. 38.
HANGING CRATE.

drawing out of the cells beyond the width of the sections. These crates are used early in the season, near the brood nest, to secure "bait" sections for the first crates, so as to induce the bees to occupy the latter **(266)**. Towards the close of the season, unfinished sections, taken from the upper crates, may be given below in these frames, to be completed **(269)**.

APPLIANCES FOR SUPERING.

Fig. 39.
SUPER BOX.

108. The Super Box (Fig. 39) is used for holding frames above the body box, or brood chamber. It is a bottomless box, the same width internally as the body box, but varying in length according to the number, and in depth according to the depth of the frames to be used in it. If for standard frames **(97)** it should be 9″ deep; if for shallow frames **(98)**, of 5½″ depth, it should be 6″ deep.

The sides, to carry the frames, are chamfered in the same way as are the inner walls of the brood chamber **(86)** and are ½″ shallower than the ends. Two pieces, 2″ × ⅝″, rabbeted 1″ × ¼″, are nailed, one on each side, their upper edges being level with the tops of the ends of the box. These pieces enclose the ends of the top bars of the frames, preventing them from shifting; they conserve the heat, and are useful also as handles.

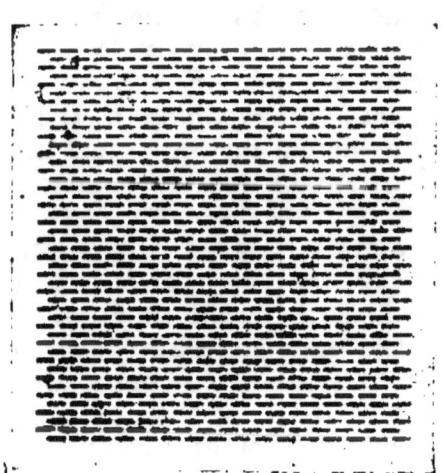

Fig. 40.
EXCLUDER.

109. The Excluder (Fig. 40) is used to prevent the queen from reaching, and depositing eggs in the sections or frames placed above the brood chamber, and for shutting off the queen and drones from any part of the hive in which their presence is not desired. **(95)**. It is a sheet of zinc large enough to cover the tops of the frames, when used to exclude from the upper storey, and perforated with holes which permit worker bees to pass, but exclude the queen and drones. In practice it is found to offer some obstruction to the workers, and it is rapidly falling into disuse among bee-keepers who work for section honey. But it has still its admirers among experienced apiarists, who claim that the advantage of having the sections protected from the queen's attentions, more than compensates for any possible obstruction to the workers **(268)**. Where super boxes **(108)** with frames are employed for extracting purposes, an excluder below the super box is generally used. It is important to arrange the excluder with its openings running across, and not parallel with, the frames, thus giving the bees freer access to the supers. New excluders may be rubbed with emery cloth, to remove the rough edges of the openings.

CHAPTER X.

COMB FOUNDATION.

110. Use of Foundation.—No less important than the introduction of the moveable frame, the invention of foundation (Fig. 41) marked a distinct advance in the methods of practical bee-keeping; simplified the management of frame hives; and effected a substantial economy in the expenses of working. It has already been pointed out that the great advantage of the modern moveable-comb hive depends upon its frames, in use, being really moveable. If bees are placed in hives fitted with empty frames, they will build their combs in the frames, but at such angles, and in such manner as frequently to fasten the frames together and to render them immoveable in the hive, thus defeating the object in view. If strips of wax, as "starters," be fixed below the top bars of the frames, the bees will begin their combs at the starters, but will sometimes build them so irregularly that, here and there, comb will be joined to comb, and only a few, if any of the combs will be perfectly even and moveable. In both cases there will be constructed so large a proportion of drone cells that the drones reared in such cells may be sufficiently numerous to consume the surplus honey which it is the aim of the bee-keeper to secure for himself **(195)**. To obviate those difficulties; to enable the

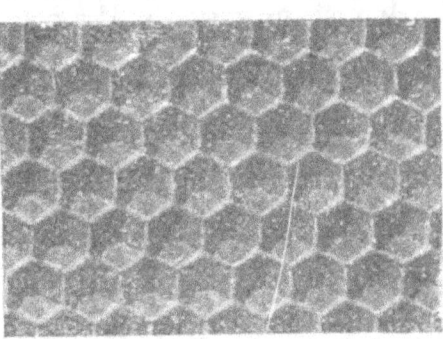

Fig. 41.
COMB FOUNDATION.

bee-keeper to exercise complete control over the work in the hive; and to constitute apiculture as a remunerative occupation, it was necessary that some means should be devised to compel the bees (1) to build straight, separate combs, hanging evenly and parallel, each within its own frame; (2) to construct such cells, worker or drone, and in such proportion, as the owner may desire; and (3) to apply to the manufacture of new combs, wax which had been used for the same purpose again and again, with "cappings" (72) and odd scraps which, otherwise, might be wasted or sold below their real value, thus preventing an extravagant consumption of honey for the secretion of wax, and an extravagant waste of time on the part of the bees during the processes of wax secretion and comb building. (73).

111. Invention of Foundation.—The application by Langstroth, in 1851, of the moveable frame principle (80) made the construction of suitable combs more than ever necessary; and, six years later (1857), Mehring, a German, of Frankenthal, produced a sheet of wax on which the shape of cells was stamped, and which was to serve as a "foundation" for the bees to build upon. Improvements upon Mehring's invention were designed to form upon the foundation the beginnings of the cell walls; and, in 1876, A. I. Root, of Medina, Ohio, U.S.A., had constructed a roller mill with embossed cylinders capable of turning out foundation in continuous sheets, and with the formation of the cells, as it is now produced. E. B. Weed subsequently devised the rolls which impress the foundation that is called by his name. These rolls are faced with type heads, and give absolute similarity throughout the sheets.

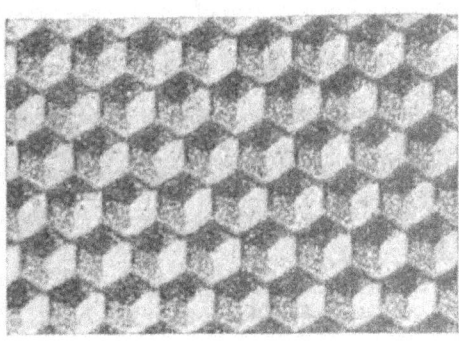

Fig. 42
SUPER FOUNDATION.

112. Varieties of Foundation.—Foundation is now supplied of various sizes, both of sheets and cells, and of various thicknesses. "Medium brood," and "Thin brood" ("Weed"), in sheets to fit the standard frame, have eight sheets and eleven sheets respectively to the pound weight, and are made both with worker cells and drone cells. "Thin super" (Fig. 42), and "Extra thin super" ("Weed"), in sheets to fill three sections each, have twenty-eight to thirty-two and thirty to thirty-six sheets respectively to the pound weight. Brood foundation is used in frames in the brood nest and super box. Super foundation is used in sections, and is made thin enough to avoid, as far as possible, the unpleasantness of a heavy mid-rib in comb which is intended to be eaten. **(101).**

113. Advantages of Foundation.—The advantages secured to bee-keepers by the use of foundation are many:—(1) When whole sheets are used in frames and sections, the combs built upon them are perfectly straight, so that they can be moved about in the hive **(110)** and transferred from hive to hive, or from crate to crate as required. (2) The combs built upon worker foundation are composed generally of worker cells, so that by the employment of this kind of foundation, the rearing of drones can be limited **(195)**, which is always a useful power in the hands of the bee-keeper. (3) All the wax produced; all old combs, scraps, and cappings removed for extracting purposes may be given back to the bees, in the shape of foundation, thus effecting a very considerable economy both of wax and time. For example—Eleven sheets of brood foundation to fill the eleven frames of a standard hive, and to supply sufficient wax for the construction of the combs, may be purchased for, say, 3s. 6d.; but it is calculated that to manufacture eleven such combs without the aid of foundation, about 13s. **worth of honey may be consumed by the bees (73).** If now we estimate the loss of honey left ungathered by the bees while secreting wax and building the bases of the cells, in the season, at 10s. to £1 we find that a supply of 3s. 6d. worth of foundation will not only greatly expedite the labours of the bees and reduce the tax upon their strength imposed by the secreting of wax, but will also effect a saving of from £1 0s. 0d. to £1 10s. 0d. per hive:—

	£	s.	d.
Honey consumed in the secreting of 2 lbs. of wax to form 11 combs, say 13 lbs. honey to the 1 lb. wax = 26 lbs. honey at, say, 6d. per lb.	0	13	0
Honey left ungathered by wax-secreting and building bees during partial construction of 11 combs, say 41 lbs. at 6d. per lb. ...	1	0	6
	1	13	6
Cost of 11 sheets brood foundation, say ...	0	3	6
Estimated saving per hive ...	£1	10	0

(These are not War prices).

Allowing for any over-estimation, if there be such, in the above calculation, there yet remains sufficient margin to point the great desirability, from a pecuniary point of view, of a generous use of foundation in the hives, and for its use in full sheets in both frames and sections.

114. Adulteration.—It should be stated here that foundation, in common with so many other articles of commerce, has not escaped the attention of the adulterator; and that it is very necessary to see that the foundation used in the hive is pure. That which is adulterated with paraffin wax, or with ordinary grease or fat, will often be refused by the bees, or if built upon, will lack the strength to endure the heat of the hive in summer, and will stretch and break down, in either case imposing much trouble and loss upon the bee-keeper. Foundation may be tested for adulteration with tallow, by the smell when broken; and for adulteration with mineral wax, by chewing for a few minutes, when if it be pure, it will crumble in the mouth, and if adulterated with paraffin or ceresin (64), will adhere in mass, like chewing gum. This test is, however, not always reliable, and some more accurate test is necessary to enable every bee-keeper to prove for himself the quality of the foundation which he buys. If a tumbler, wine glass, or wide-mouthed bottle be half filled with water, and a small piece of pure wax, such as may generally be found somewhere in a hive, be dropped into the water, it will float, because the specific gravity of the wax (960-970) is less than the specific gravity of the water (1,000). If now a small quantity of alcohol be slowly poured into the vessel until the piece of wax no longer floats, but just sinks to the bottom, and no more alcohol be added than that which is just sufficient to permit the wax to descend from the surface, then the water will have been brought to the same specific gravity as that of pure wax. But wax that is adulterated with paraffin or ceresin is of lighter specific gravity than that of pure beeswax, and if dropped into the vessel it will float where pure beeswax will not. This test is inexpensive, and sufficiently accurate to serve for practical purposes in the examination of foundation for adulteration with mineral wax. The liquid may be kept in a glass-stoppered bottle for future tests.

115. Change of Colour.—When no longer fresh, foundation may become darker in colour and so brittle that it will break if tested by bending. Warming it slightly before a fire will improve it, and will partially restore its original colour.

116. Quantity Required.—1¼ lbs. of brood foundation will about suffice for 11 standard frames (97), 1⅛ lb. of thin super foundation will fill about 105 sections, 4¼″×4¼″, or five crates (103).

117. Fixing Foundation.—Foundation is usually fixed in sections by means of the splits in the tops, or in the tops and sides **(101)**, and in frames, the upper edge of the sheet is caught either in a saw-cut in the top bar, or by the groove and wedge already described **(97)**. When full sheets are used in frames, and especially when intended for extracting purposes, the foundation is generally wired to the frames **(263)**. Formerly foundation was fixed with melted wax, but this method, requiring more time and labour, is rapidly falling into disuse. However the fastening may be made, it is important that the foundation be fixed right side up (Figs. 41, 42, 43).

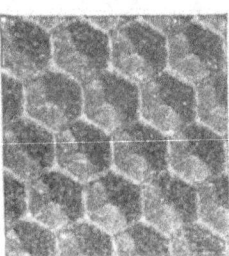 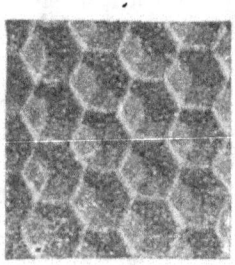

Fig. 43
FOUNDATION.
A, RIGHT.   B, WRONG.

It has been observed by Huber, and Cheshire, and indeed by everyone who has carefully examined a honey comb, that it is customary for bees to build their cells with two of the six sides perpendicular, thus— and in this position foundation should always be used (Fig. 43, A). If the sheet be fixed the other way, the impress of the cells will be out of form, thus— and this is not desirable (Fig. 43, B).

113. Wiring Appliances are used for fastening foundation securely in frames to prevent it from sagging when the heat of the hive softens the wax, and the weight of clustering bees tends to bear it down. Combs which may some day find their way to the extractor **(134)** should always be wired in the frames, lest the centrifugal force employed to throw out the honey should break the comb. The Wiring Board (Fig. 44) is a piece of $\frac{3}{8}''$ wood, cut $13\frac{1}{4}'' \times 7\frac{1}{4}''$ so as to fit inside the frame. Two laths, $14\frac{3}{4}'' \times \frac{7}{8}'' \times \frac{3}{8}''$, projecting $\frac{3}{4}''$ at each end, are nailed on the back. Holes are bored in the bars of the frame; and No. 30 tinned

Fig. 44
WIRING BOARD.

wire is drawn through and tightened **(262)**. It is then embedded in the foundation by a heated embedder, which may be a strong bradawl, having a groove cut in its edge. Drawn along the wire it presses it into the foundation, at the same time melting sufficient wax to cover the wire **(263)**.

COMB FOUNDATION. 69

The Woiblet Spur-embedder (Fig. 45) has a grooved wheel to act upon the wire

Fig. 45.

WOIBLET SPUR-EMBEDDER.

The following method of embedding the wire by electricity may be of interest. The frames are wired as directed (**262**), but a short length of wire is left when cutting off after twisting round the tacks. The foundation is next fixed and the frame thus prepared is laid on the wiring board with the wires underneath. Two wires are now taken from the terminals of a 4-volt accumulator, such as is used for ignition purposes in motor-cars, etc. One of these wires is hooked into the wire on the frame where it passes on the outside of the side piece between B and C; the other wire from battery is now held with

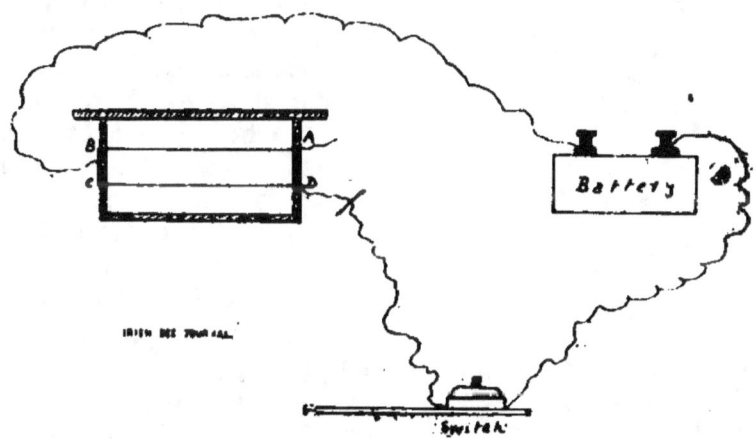

Fig. 45b.

ELECTRIC EMBEDDER.

one hand in contact with one of the loose ends of frame wire (A or D). The current of electricity, which will now pass through the portion of the frame wire connecting the battery wires, heats this portion, and a gentle pressure on the foundation with the free hand causes the wire to sink into the wax, leaving behind scarce a mark to show where it entered. When the wire shows through the bottoms of the cells on the upper side, the battery wire is freed from contact with the frame wire at tack, and the other strand is treated similarly. The putting into circuit of a switch, which can be pressed by the foot, greatly facilitates the operation by leaving both hands free. This switch can easily be made by mounting an ordinary bell-push on a small block of wood.

CHAPTER XI.

APPLIANCES FOR FEEDING BEES.

119. Feeding.—Bees require to be fed when their stores run short; and at other times, also, it is found to be profitable to supply artificial food **(311)**. For this purpose, it is necessary to have feeders which will supply the food in the proper quantities, and in the proper position, so that the bees may use it for the purpose intended, and may have convenient access to it, without the danger of setting up robbing **(307)** by attracting stranger bees to the sweets supplied.

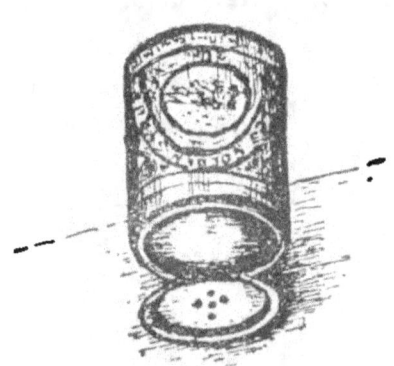

Fig. 46.
ECONOMIC FEEDER.

120. The "Economic" Feeder (Fig. 46) is an ordinary syrup tin, with a lever-top lid in which holes are punched. It is inverted upon the frames direct, or upon a single stage of ¼" wood through which a hole has been cut to give access to the bees. When the feeder is being removed, a corner of the carbolic cloth **(127)**, or a separator **(102)**, may be slipped under it to keep the bees down.

Fig. 47.
BOTTLE AND STAGE FEEDER.

121. The Bottle and Stage Feeder (Fig. 47) can be put together at a trifling expense. It consists of a wide-mouthed bottle, or jar, with a piece of coarse calico tied over the mouth, and two squares of ¼" wood. In one square a round opening is cut large enough to admit the mouth of the bottle, and in the other square an opening ½", or 1", smaller in diameter. The squares are then nailed evenly one upon the other. This stage is placed upon the tops of the frames. When the bottle is being removed, a piece of zinc, or cardboard, may be slipped between it and the opening in the stage, to prevent the bees from escaping upwards.

APPLIANCES FOR FEEDING BEES.

Fig. 48.
GRADUATED FEEDER.

122. The Graduated Bottle and Stage Feeder (Fig. 48) is made upon the same principles with those already described, but the screw-cap and the stage are so arranged that by turning the bottle round, the supply of syrup can be increased or diminished between 1 and 9 holes, or can be cut off altogether. A pointer attached to the screw-cap, and figures upon the stage indicating the number of holes exposed, enable the supply to be regulated as desired. This feeder can be used for slow or rapid feeding according to the season.

123. Slow and Rapid Feeders (Figs. 49, 50), capable of holding 1 quart of syrup, are used chiefly in the autumn when it becomes necessary to feed up the stocks rapidly so that they may be able to store and seal the syrup before the cold weather sets in--wintering bees upon unsealed stores being very likely to lead to dysentery **(330)**. The feeder (Fig. 49) is a round tin box, with a moveable lid, and a flange round the bottom to provide the necessary bee space between the frames and the feeder when in use. A round hole in the bottom permits the bees to pass up a funnel into the feeder. A wooden float surrounds the funnel, and outside this is a tin case with a glass top. When the lid is removed, and syrup is poured into the box, the wooden float rises. The bees pass out upon the float to reach the syrup, and can be seen through the glass top of the inner case. The "Alexander" Feeder (Fig. 50), described and illustrated in *Gleanings*, is attached to the floor board, and the hive is drawn back to cover it **(91)**. All that is necessary is to lift a block off the projecting end, pour in syrup, and replace the block.

Fig. 49.
ROUND TIN FEEDER.

Fig. 50.
"ALEXANDER" FEEDER.

124. The Canadian Feeder (Fig. 51), capable of holding six to ten pounds of syrup, is used when it is desired to give food rapidly, or to have the winter food for a number of colonies stored and sealed by one stock **(315)**. In the latter case the stock is supplied with drawn out combs, and the feeder is refilled as fast as it is emptied, the combs being removed when sealed, and their places supplied by empty combs. By setting apart a stock for this purpose sufficient sealed stores can be provided to supply all the colonies with winter food. The feeder has a tin lining, and is fitted with a wooden construction to give the bees foot-hold. This latter can be removed when it is desired to insert honey in comb, either for feeding or for cleaning up purposes. There is a double-hung lid, so that the contents can be seen, and the feeder replenished as required.

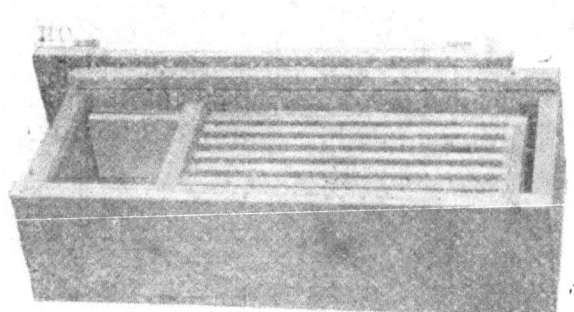

Fig. 51.
CANADIAN FEEDER.

125. The Division Board Feeder (Fig. 52) is a device for giving food in the body of the hive. It is made the same length and depth as an ordinary Division Board or Dummy **(93)**. The top bar is fastened with screws so that it may be removed for cleaning purposes. The food is poured through a hole in the top bar, close to which hole a partition running from within ¼" of the bottom to the top shuts the bees off from the hole. A ¼" slit in one side, near the top, admits the bees to the syrup.

Fig. 52.
DIVISION BOARD FEEDER.

CHAPTER XII.

APPLIANCES FOR SUBDUING AND HANDLING BEES.

126. The Smoker (Fig. 53) is employed for subduing bees, and is a most useful appliance in an apiary **(171)**. A puff or two of smoke blown in at the entrance frightens the bees, and causes them to fill themselves with honey; in which condition they are not inclined to give trouble **(167)**. The smoker has a bellows, a fuel box, and a removable nozzle. A roll of dry brown paper, a piece of rag, or a piece of dry, rotten wood is lighted and placed in the fuel box, lighted end down; the nozzle is put on, and if the smoker be left standing nozzle up, it will draw like a chimney, and the fuel will keep alight. To extinguish the fuel the smoker is placed on its side. If brown paper is used it may first be damped with a weak solution of saltpetre, and then thoroughly dried **(175)**. A grating in the lower part of the nozzle prevents the blowing into the hive of particles of lighted fuel. Smokers require to be cleaned occasionally. The nozzle may be cleaned by being boiled in water. If the fuel box require cleaning, it may be similarly treated, after having been removed from the bellows.

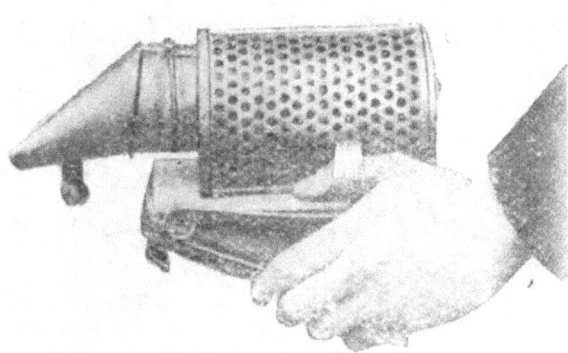

Fig. 53.
SMOKER.

127. The Carbolic Cloth is also a subduer of bees, and by some is preferred to the smoker. In certain operations it is somewhat easier to work with than is the smoker, and once prepared **(176)**, it requires little or no attention during a long period. Ticking, calico, or muslin, 20″ × 18″, may be used, with (if preferred) a hem on one 18″ edge to take an 18″ lath. A solution of Calvert's No. 5 Carbolic Acid, one part to ten parts of water, is prepared, and with it the cloth is thoroughly

Photo by] *[J. G. Digges.*
IRISH BEE-KEEPERS' ASSOCIATION'S BEE TENT. EXAMINING CANDIDATES FOR EXPERT CERTIFICATES.

saturated. If the cloth, when not in use, be kept in a close-shut tin box, it will retain its objectionable smell for a long time. The solution should be shaken before being used. (Recipe 362).

128. Use of Veils.—The veil (Fig. 54) is used to protect the face and neck from stings. Although bees in a hive may be thoroughly subdued by smoke or carbolic fumes, an occasional bee outside the hive, and which has not been within reach of the subduing agent, may develop a warlike spirit sufficiently active to be taken account of (169). Many bee-keepers, from oft familiarity, hold stings in contempt, no matter where applied, becoming immune to the poison when thoroughly inoculated with it. But others, and especially beginners, are wise in having veils for the protection of the face, and because of the confidence they give during the manipulation of unamiable stocks. The veil may be made of black netting, or of white netting if a piece of black be added for the front, it being easier to see through black netting than through white. A piece of netting 48" × 24" will make an ample veil; and 36" × 18" will make a veil sufficiently large for most purposes. The ends are sewn together, and a hem is run on one edge, to carry a piece of elastic arranged to grip tightly round the crown of the hat to be used. If worn as in the illustration (Fig. 54), the free end being carefully tucked in and the coat buttoned, it will be next to impossible for the face to be stung. Some veils have a piece of elastic round the lower edge also as an additional protection; it closes round the collar, effectually preventing bees from crawling under.

Fig. 54.
NET VEIL.

129. A Lady's Veil may be made larger, for wear with a broad-brimmed hat. A strip of broad elastic is sewn in the lower edge so that it will fit over the shoulders, and two straps passing under the arms and buttoning in front, keep the veil in position. Ladies sometimes wear overalls to which the veil fastens, and which are tied or strapped round the waist (169). Mosquito netting makes an excellent lady's veil.

130. The Wire-cloth Veil (Fig. 55) is not so comfortable as a net veil, but it has the advantage that the wind cannot blow it against the tip of the nose or chin at the precise moment when an aggressive bee is seeking a point of attack. On the

Fig. 55.
WIRE-CLOTH VEIL.

other hand, it is not easy to carry about, and when midges are worrying one's face, it is exasperating to be unable to get at them with the fingers. Net and elastic are used round the crown of the hat, and from the rim down are two pieces of wire cloth 18" × 7", joined at the ends by two pieces of netting 4" × 7", allowing the veil to fold flat when not in use. The veil is carried down a few inches more with netting, which may be either tucked under the coat or caught round the collar by elastic.

131. Use of Gloves.—Among experienced bee-keepers gloves are held in ill-repute, as clumsy and unnecessary things. Some will almost go so far as to hold that nothing that is accomplished with the aid of gloves can properly be called bee-keeping. During many years it was impossible for anyone who covered his hands to qualify as an Expert under the Rules of the Irish Bee-keepers' Association, and, quite wisely, under the existing Rules, a candidate for the Association's Expert Certificate is penalised at his examination, if he resorts to the use of gloves during manipulations. Nevertheless, gloves, tho' they be "clumsy things," are in some cases indispensable; and, tho' they be sneered at by veterans, are often a source of confidence, and, as such, a valuable assistance to beginners (169).

"There are constitutions that cannot endure stings without much pain and inconvenience, and occasionally even positive danger. There are ladies, and strong men also, to whom a 100 per cent. hypodermic injection of formic acid, accompanied by the angry buzz of a vicious bee, is alway a hateful experience; and there are many who will agree as to the clumsiness of gloves, but who would not dare, without such protection, to engage in the practice of bee-keeping at all."—*Irish Bee Journal.*

It would be absurd to dissuade such persons from bee-keeping for no other reason than that they manipulate with gloves. It must, however, be said that some of the most delicate operations, such as picking a queen off the comb, and wing clipping (212), cannot be carried out as neatly, nor always as successfully, with gloves; and that, generally speaking, the wiser course is to discontinue the use of gloves as soon as possible.

132. Various Gloves.—Thick woollen gloves, covered with a pair of white cotton gloves long enough in the wrist to run up on the cuff, are sometimes used; but they are really "clumsy things," and are difficult to work with **(169)**. If soaked in water immediately beforehand, the bees will not be inclined to sting them; and if stung, their thickness prevents the sting from entering the flesh. India rubber gloves are good protectors, and are not inconvenient to work with. Ordinary leather gloves may, sometimes, be made to serve the purpose. They are steeped for a minute or two in hot water; are then put upon the hands; and while being held before a fire, have beeswax well rubbed into the leather. It is said that bees commonly respect such gloves, and that if the tips of a finger and thumb be removed from both gloves, and the exposed flesh be anointed with wax, all manipulations can be carried out with safety. Burkitt Bee Gloves **(169)** are made of soft white leather, having attached a linen gauntlet coming well up the arm, and bordered with red braid. These gloves are quite ornamental when new; the bees do not show any inclination to attack them; and operations can be performed without inconvenience.

Photo by: [*J. G. Digges.*
EXAMINING THE TOP CRATE—A CIGARETTE "SMOKER."

CHAPTER XIII.

APPLIANCES FOR HONEY AND WAX EXTRACTION.

133. Invention of the Honey Extractor.—When Langstroth, in 1851, had invented the moveable-comb hive **(80)**, and Mehring, in 1857, had introduced foundation **(111)**, the next great benefit which discovery was to confer upon the industry was the provision of a means by which honey might be removed from the comb without the destruction of the latter. Hitherto honey had been extracted either by crushing the combs or by melting them—an expensive method in every way, as will be gathered from what has already been said upon the subjects of wax secretion and comb building **(73)**. In 1865, de Hruschka, of Venice, observing his son carelessly swinging a piece of honey comb in a basket, noticed that the motion slung some of the honey out of the cells. Taught by what seemed to be a mere accident, he proceeded to apply the principle of centrifugal force to honey extraction, and with a diligence which was crowned with success, and which has for ever placed bee-keepers under a debt of gratitude to the man

Fig. 56.
HONEY EXTRACTOR AND KNIVES.

134. The Honey Extractor.—The Honey Extractor is a strong, tinned iron can (Fig. 56) with two cages, which revolve round a verticle spindle, and hold each a frame of comb. The cages (*b*) are set in motion by the handle (*a*) on top, and when the honey has been thrown from the outer sides, the combs are reversed and the operation is repeated. The honey is slung out against the sides of the can, is received in the bottom, below the revolving cages, and may be drawn off through the syrup tap (*d*) **(277)**. If the extracting is properly done, and if the combs have been wired in the frames **(263)**, the combs remain uninjured, and may be returned to the hive to be refilled, being used for this purpose year after year, thus effecting a great economy, as already explained **(73).** Extractors are made with gearing, which lessens the labour of working the revolving cages, and is capable of getting up a high speed. A small form of extractor may be had to extract one frame at a time. This appliance, however, is slow, laborious, and not likely ever to become popular.

Fig. 57.
UNCAPPING KNIFE.

135. Uncapping Knife.—Uncapping may be done with a sharp carving knife. But the most useful knife for the purpose is the uncapping knife illustrated. (Figs. 56, 57). It has bevelled edges; and with a little practice it can be used with rapidity and completeness upon the most irregularly built combs **(276).**

Fig. 58.
STRAINER AND RIPENER.

Fig. 59.
RYMER HONEY PRESS.

136. Strainer and Ripener.—Honey that has not been capped over by the bees before extraction, and that is unripe, requires to be ripened in a warm temperature before being offered for sale. The Strainer and Ripener (Fig. 58) has a flannel, a cheese cloth, or a wire-gauze strainer, into which the honey is run from the extractor. The honey flows into the ripener underneath, and a syrup tap is provided for drawing it off. **(277).**

137. The Honey Press.—Heather honey, which is too thick to be thrown out by the Honey Extractor **(134)**, and honey which is to be removed from combs that are intended to be rendered into wax, may be pressed out by the Honey Press, or, in small quantities, by a potato masher. The Rymer Honey Press (Fig. 59) is made of malleable iron and steel; it has a square thread screw, and all the parts that come in contact with honey are tinned. The honey is forced between the grate and the outer case, and flows into the receptacle underneath **(276)**.

138. Wax Extractors. — These most useful appliances are intended for the rendering into wax of discarded combs, cell cappings **(276)** and any odd bits of foundation which may be collected from time to time, and (wax being a valuable commodity) may thus be turned to good account **(279)**.

Fig. 60.
SOLAR WAX EXTRACTOR.

139. The Solar Wax Extractor (Fig. 60) is simple in use, inexpensive, and gives satisfactory results, provided that the solar element be not wanting. Mr. M. H. Read describes his home-made extractor (Fig. 61) as follows:—

"The extractor measures 2′.3″ long, 1′.11″ deep, 10″ high at back, and 6½″ high in front, inside measurement. It is made of 1½″ timber, dovetailed. The sides and back of the sash, or cover, are 2″ by

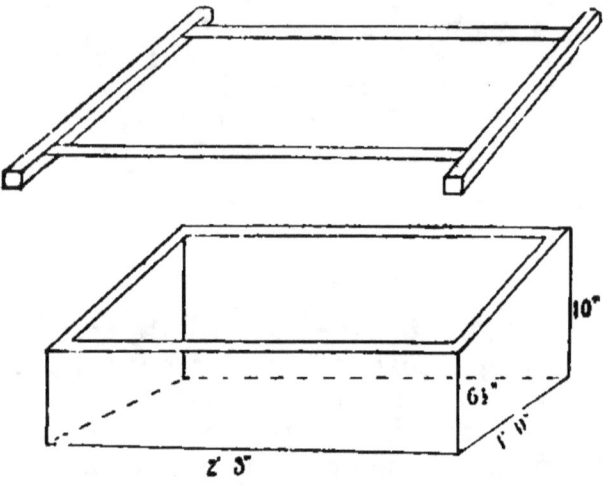

Fig. 61.
READ'S SOLAR WAX EXTRACTOR.

1½″, and the front is 2″ by 1″. The sash is glazed with 24oz. glass. The inside width (2′·3″) was fixed upon so as to hold a tray of ordinary corrugated iron, upon which the wax to be purified is laid. The impurities remain on the tray, and the clear wax, as melted, runs down into a trough which runs all along the front. The trough is cut from a strip of corrugated iron (valley and two corrugations) the two corrugations hammered up, to make one deep trough, the ends of which are hammered up, so that the wax is held in it. The extractor has a loose-fitting bottom, and is filled up with a sloping bed of cinders, within 4″ of the glass, the trough being set in the ashes in front, and the tray on the bed of ashes and overhanging the edge of the trough. Being fitted with a loose bottom, the extractor can be turned to face the sun, the rays of which should fall perpendicularly on the glass. A reflector, of a large sheet of tin, could be added."—*Irish Bee Journal*.

Fig. 62.
STEAM WAX EXTRACTOR.

140. Steam Wax Extractor.—For extracting by steam, an apparatus (Fig. 62) is supplied which does its work thoroughly, is not difficult to manage, and in our fickle climate which so often denies us the sun, is more generally useful than is the Solar Extractor. The upper portion has a perforated tin basket into which the pieces of comb are placed, and underneath which is a tray, with an outlet. When the water in the lower portion boils, the steam ascends to the basket and melts the wax which escapes through the outlet and is caught in a basin of cold water. The refuse remains in the basket, and when the wax in the basin cools, it may be lifted out in a cake. (279).

[*Other useful and necessary Appliances will be found described in the various chapters of Part III.*]

PART III.

MODERN BEE-KEEPING.

CHAPTER XIV.

PAST AND PRESENT.

141. Past Ignorance.—Previous to the introduction of what are known as "modern methods," bee-keeping was carried on under most discouraging conditions. It is true that in very early days something was known of the habits of the honey bee, and that so early as 70 B.C. Virgil, the Latin poet, put forward in verse the results of his study of the habits of bees, with a degree of accuracy sufficient to excite a wondering admiration on the part of twentieth century readers. But to the average bee-keeper the hive, until comparatively recent times, was as a sealed book; and the marvels that it contains, the excellence of its internal economy, and the unselfish devotion, wisdom, and singular attractiveness of its occupants were, if known at all, known only to the few. Virtues and beauties thus hidden could make but little appeal, as yet, to the respectful admiration of human intelligence. The heroic acts and incomparable works were wrought, like evil deeds, in darkness: and man, loving only the visible, the tangible, sceptic always of the unseen, had not learned that within the secret places of the hive were enshrined mystery upon mystery, and that within the humming insect, flitting in his garden from flower to flower, there beat a heart brave and noble enough to deserve his respect and even to awaken his love.

142. Survival of the Unfit.—Unfamiliar with the instincts of bee life, man found himself unable to control by gentleness, and thought it necessary to resort to violence for the subjugation of insects armed by nature with stings. The harvest of honey and wax was gathered at the expense of the lives of the colonies. The strongest and fittest—those whose stores were heaviest, were devoted to destruction; the weakest and the sickly were spared; and the sulphur pit—that abominable outrage upon industrious innocence, laid waste the home of vigour and opulence, and secured the survival of the unfit (77).

It remained for Swammerdam, Reaumur, Huber (80) and other investigators, to dispel the darkness which surrounded the operations of the hive, and to devise means by which the occupants might be controlled, and the industry be worked for increased profit, and upon humane principles. Nor was it the least important result of their researches which put it within our power to correct the errors of the past; and, by careful selection, to effect such improvements in the race of bees as may tend to render them more robust, less liable to disease, gentler, and more prolific and profitable.

143. Modern Bee-keeping.—The title "Modern Bee-keeping" stands for such skilful management of bees, based upon an intelligent appreciation of their habits, as may secure the maximum results of their labours, and the fullest development of their best characteristics. It represents the desire to minister to their comfort; to assist their industry by thoughtful anticipation of their requirements; and to encourage in them the spirit of amiability by the display of a like spirit on the part of man, and by the avoidance of all roughness and cruelty in dealing with them. Modern bee-keeping has so improved upon the older methods that the produce of the bees' labour has been enormously increased, without a corresponding tax upon their strength. It has been made possible for anyone who understands what it means to take pains, to manage bees with a handsome profit to himself. He can now engage in a pursuit which has in itself an enthralling interest; and which, if carefully attended to, will return more than an ample compensation for the time devoted to it. The moveable-comb hive permits him to become familiar with the habits, and to explore all the wondrous work of the honey bee (81). He can take out the "waxen palaces," can investigate their beauties, and see with what skill they have been constructed. He can watch the queen as she moves across the combs, depositing her eggs in the vacant cells. He can replace her with a younger queen reared by himself, or imported in a postal packet from foreign lands (300), and can, at will, oblige her to produce drones or workers (113), as the conditions of his colonies require. He can observe the various stages of the egg and larva, and witness the breaking of the capping and the emerging of the new-born bee. By the use of foundation (110) he can supply the material for the building of the combs; and can so regulate the storing of honey that it may be removed in the shape and condition most marketable, and without injury to the gatherers. The extractor (134) enables him to use the same combs again and again, and, by increasing the harvest, to make his industry still more profitable. In short, he can so utilise the advantages which modern discovery

78   THE PRACTICAL BEE GUIDE.

AN OUTBREAK OF "BEE FEVER" AT KILTYCLOGHER.
(1) A Beginner's Apiary. (2) Hive Makers. (3) His First Swarm.

and invention have supplied, that he can engage in bee-keeping as in a delightful occupation, and one that is capable of being turned to good practical account.

144. A Profitable Industry.—It is something in favour of modern bee-keeping to be able to say that, in proportion with the amount of labour and capital involved, no other agricultural industry can show a like profit. A good stock of bees in a modern hive, with the necessary fittings, and costing in all, say £1 10s. is capable of producing, in a normal season, and under proper management, a profit of from £1 10s. to £2, or cent. per cent., and over. It is something also in its favour that it requires neither broad acres nor much physical strength for its employment. Four square feet of land will hold a hive. A window sill will accommodate two. A corner of a yard or garden—a plot 25 feet square, might be occupied by from 25 to 50. There is no heavy labour required. For five or six months of the year there is little to be done. In the remaining months an average of a quarter of an hour per week should suffice to devote to one stock. And it is open-air work, light, interesting, and such as ladies, and even children can accomplish without fatigue. There are many school girls and boys who are working bees successfully, and are making handsome additions to the family purse. There are bee-keepers not a few who, without excessive labour, are marketing over £200 worth, each, of honey per annum. There is the case of Mr. John Doyle, of Kellystown—a case which offers a sufficient reply to folk who sneer at the industry as a mere hobby. Starting in 1887 with the discovery of a stray swarm, he had found bee-keeping increasingly remunerative, and had so applied the modern principles to his industry that in 1901 his bees paid him a profit of over £100. In 1906, he marketed over £166 worth of honey from 92 stocks, and from the profits produced by his bees he acquired land and houses, his latest purchase—Woodville House and farm, having cost £1,000.— (*Irish Bee Journal*). Nothing of the kind could have been possible under the old methods. Bees in skeps or boxes cannot produce anything like the same profit which they are capable of producing in modern hives and under capable management. It is surprising that it should still be possible to find whole districts in which modern bee-keeping is unknown, and where only the wasteful, cruel methods of the skeppist are practised.

CHAPTER XV.

ARRANGING AN APIARY.

145. Selecting a Position.—Before actually beginning bee-keeping, it will be well to select a suitable position for the apiary. Bees in hives are sometimes kept in curious places—on house roofs; in narrow passages; on window sills. A lady in London has several stocks in her drawingroom. A hive may be set up in one's bedroom, the bees having a passage through a hole in the window sash. For an apiary out of doors almost any position will suit. But there ought not to be any serious obstruction to the bees' flight; and there should be room at the back of the hives for the owner, and a reasonable distance between the apiary and the county road or other place of public resort.

146. Bees near Dwellings.—It can hardly be said that bees learn to know their owner as a dog learns to know his master: yet it has been observed that bees located near dwellings become accustomed to persons passing to and fro, and are less likely to make themselves objectionable when one approaches their hives, than if they were situated in a remote, quiet place. Indeed bees, remarkable at one time for their gentleness, have been known to develop very hasty tempers after having been removed from their old stand near a dwelling, to a lonely spot where they were never visited except for the purpose of manipulation. Risk of unpleasantness may be minimised by a wise arrangement of the stocks. If, for example, the hives be placed thirty or forty yards south of a dwelling, and with their backs to the house, the flight will be towards the south, and the bees will give little or no annoyance. Should there be a path or garden in front of them, a high fence, or hedge, will "lift" them over the path, and will serve as a protection for persons passing by. It is desirable to provide against cold storms from the north and west. A hedge of strong privet plants will quickly make an efficient shelter. Although not absolutely necessary, it is advisable to have the hives, or as many of them as possible, facing south-east; because, in that position, they will get the warmth of the early sun about their entrances to entice the occupants out for early labour.

147. Position of the Hives.—The hives ought not to be crowded together. Bees, on taking flight, mark the location of their hives, and with surprising accuracy return from long distances to the same spot from which they started **(156)**. But, when their hives are close together, and are painted the same colour, with no distinguishing marks upon them, bees will sometimes enter the wrong hives and meet a warm reception there, leading to fighting and general excitement, which should be avoided as far as possible. Bees of a colony quickly detect an intruder. In the case of queens returning from their nuptials, it is of the first importance that they should have every facility for recognising their own hives **(283)**. The hives may stand four feet from each other, or farther apart if space permit.

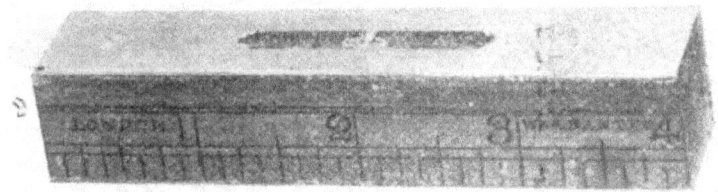

Fig. 63.
SPIRIT LEVEL AND MEASURE.

Those illustrated are six feet apart and are painted (1) white, (2) red, and (3) blue, in succession. They should be perfectly level across the frames in order that the combs may be built plumb. If the frames run at right angles to the entrance, it is no harm to give the hive a tilt towards the front. Hives with legs may be set upon bases of flag or concrete, upon four bricks, or directly upon the ground. If bases or bricks are used, they should be carefully levelled with a spirit level (Fig. 63). A hive without legs (Fig. 65) may have a simple stand made of two pieces of plank 24" × 7" × 1½", a piece 3½" × 1½" being cut out of each, and the pieces being nailed together in shape of an X (Fig. 64). The floor board **(85)** being placed in position, should be tested with a level, or with a bowl of water set upon it. When the hive is ready a stout stake may be driven

Fig. 64.
"X" STAND.

into the ground, at one side of the hive, for use when preparations are being made for winter **(380)**. Grass and weeds must be kept down in the vicinity of the hives. If allowed to grow, they intercept the flight, and should a queen drop off a frame, or a clipped queen **(212)** fall on the ground at swarming time, she may easily be trodden on, or lost. It will save much trouble

Fig. 65.
"X" STAND IN USE.

Fig. 66.
APPLIANCE PRESS.

if the sod be lifted around and in front of the hives, and a good coat of concrete, gravel, or cinders be laid down. The plot selected should be fenced to prevent cattle, pigs, etc., from interfering with the hives.

148. Appliance Press and Apiary House.—It is most inconvenient to have the appliances stored at a distance from the apiary. To leave things about the house means often to have them mislaid; and to run to and fro when engaged manipulating is often to put an undue tax upon the temper, and to raise a riot among the bees. A press on legs, made of old boxes, painted, with a waterproof roof (Fig. 66), and that can be carried from place to place if necessary, will be found to amply repay the trouble of making it. For an apiary of more than five or six hives, an apiary-house should be provided in which all tools and appliances can be kept, frames and sections put together, hives nailed, honey extracted, and all the various jobs have attention. Such a house can be made at a trifling expense, while the assistance it gives, and the time it saves are incalculable.

CHAPTER XVI.

COMMENCING BEE-KEEPING.

149. Three Words of Advice may be useful to anyone who proposes to try his hand at bee-keeping, viz. :—Begin

    (1) Moderately.
    (2) Prudently.
    (3) Intelligently.

150. Begin on a Small Scale.—To begin moderately, begin on a small scale, with one or two stocks. Few things have done more to discourage beginners from persevering with the industry, than has the mistake of starting with more stocks than they could easily manage before they had gained the necessary experience. In due time, when you have learned something of the habits and wants of bees, you will be able to add to the number of your colonies and, perhaps, to attend to twelve or twenty stocks without greater expenditure of time than, at the outset, you will find necessary to devote to two.

151. Purchasing Bees.—To begin prudently, provide yourself with the best hives and appliances that you can get; not necessarily the most expensive, but the best. And if you start by purchasing bees, do not hesitate to give a little more money for a really good stock or swarm. By "Stock" is meant an established colony of bees in a hive. But stocks differ so much in value that if one be worth £1 another may not be worth half-a-crown, and a third may be worth less than nothing. By "Swarm" is meant a queen, and attendant bees which have just abandoned a hive **(19)**. These also vary in value from 10s. or 15s. to nothing, according to their numbers, condition of health, the ages of their queens, and the date of their swarming **(205)**. Most important is it to provide against purchasing, or admitting as a free gift to your apiary, **bees that are diseased or that have come from a diseased hive, a**

diseased apiary, or a diseased neighbourhood (349). It is safest, and often it is necessary, to get someone of experience to inspect bees about to be purchased and the apiary to which they belong, and to report upon their antecedents and condition. On this subject perhaps no one can offer advice sounder, or based upon wider experience than that given by Mr. Turlough B. O'Bryen—

"Just now the general desire to purchase stocks to make an early start, tempts me to say a word both of caution and advice on the matter. Foul brood is now so widespread that no county can be said to be absolutely free from it. Therefore the would-be purchaser should take the offer of a bargain with suspicion and caution. It will not do to trust to a piece of comb to send to our Editor, for that particular piece may not be diseased. It is better to have the stock examined by one who is familiar with the disease in all its phases. Being satisfied that the stock is healthy, it only requires a glance into the centre brood frames to ascertain if there is brood in all stages, or any in the junior stages (eggs and larvae) to certify the presence of a queen."—*Irish Bee Journal.*

152. Commencing with a Swarm.—Speaking generally, one may begin at any time of the year in which the "bee fever" takes him. But, if he can arrange it so, it will be best to commence in the spring; and, having his apiary and hives in readiness, to purchase the best early swarm that he can procure. An ideal swarm will be one that comes off in April, or early in May; that is from a stock which swarmed in the previous year; and that contains from 25,000 to 30,000 bees. If it issue early in the spring, it will be able to give some surplus honey, and to establish itself well before winter (205): if it is from a stock which swarmed in the previous year, it will have a queen in her prime (20), and if it contain 25,000 or 30,000 bees, it will be strong enough to put heart into its work, and to carry on until the new brood shall be able to fly. The vendor will probably hive the swarm in a skep (77), or box, and will deliver it, or will notify the purchaser that it is ready for removal.

153. Moving Swarms.—The transporting of the swarm to the new apiary presents no difficulty. All that is required is to remove it as soon as it has been secured, or, failing that, to wait until the bees have settled down in the evening. The skep should be set down upon a piece of perforated zinc, coarse netting, or canvas, which should be firmly tied round the skep, so as to make sure that sufficient air can get in, and that no bees can get out. The swarm, thus secure from escape, and asphyxiation, is carried, inverted, to its stand in its new home. If now the skep, or box be weighed, and if it be weighed again when empty, and if 5,000 bees be allowed to the pound, the number of bees in the swarm can be fairly accurately calculated. A lot of 3 lbs. weight, including say 15,000 bees, makes a moderate swarm; 5 lbs., or say 25,000 bees, may be regarded with very particular satisfaction. From 3s. to 3s. 6d. per lb. is not too much to pay for an early swarm with a one-year-old queen. Fig. 67 illustrates a cheap and useful balance for weighing hives.

Fig. 67.
BALANCE FOR WEIGHING HIVES, UP TO 400 LBS.

154. Sending Swarms per Post.—Swarms, and bees apart from their combs, may be sent cheaply and safely per post. A suitable box is procured, and two webb or canvas bands, 4" or 5" deep, are fastened in it parallel with and about 2" from the sides, and at least 1½" from the top and bottom. This is done by running string or stout wire through holes bored in the ends of the box, and through hems in the canvas bands. Instead of a lid, the box should have a covering of coarse, strong canvas, and, for further ventilation, several small holes should be bored in the bottom and sides. Mr. O'Bryen, from whom the idea comes, has sent bees with perfect safety in this manner. He, however, took the precaution of asking the postal officials to send the box in either a vacant, or a not overcrowded post office basket.

155. Commencing with a Stock.—If it be decided to begin by the purchase of an established stock in a skep or frame hive, steps should be taken to ascertain exactly the condition of the colony and of the skep or hive. If the vendor can show a clean bill of health; if the queen be vigorous; the combs even, and well supplied with brood, and not too old; and if the skep or hive be in good condition, a bargain may be made.

156. Moving Stocks.—The transport of a stock to a new apiary requires some care. Among other considerations, the question of distance must be taken into account. It has already been explained that bees will fly a distance of two miles in search of food, and will return again to the place from which they started (35). But if their hive be moved more than a couple of feet during their absence, or at night, without precautions being taken to cause the bees, in the morning, to take notice of the alteration, a number of them will return to the stand to which they were accustomed, and will flutter about it and die there. It follows that if the bees to be purchased are located more than two miles from the purchaser's apiary, they may be transported direct without risk; and that if their old home is less than two miles off, special care must be taken to prevent their return to it. Put briefly, bees may be moved directly two feet and under, or two miles and over; but for intermediate distances due precautions must be observed, viz.—either to move the stock by short stages of two feet, in the evening, after the bees have been flying freely, or to move it from the old stand to a spot more than two miles away, and to let the bees fly there for a few days, then moving them to their new home; provided always that this second journey be not less than two miles. If neither of these precautions can be adopted, a third one may possibly answer the purpose, viz.—the bees can be carried to their new location in the evening, and arrangements can be made to oblige them, when they fly in the morning, to observe that the scene, meanwhile, has changed. This is done by placing about the entrance such impediments to their flight as will demand their attention. A little grass may be pushed into the entrance so that they may have to squeeze their way out; a cloth may be hung down in front of the hive; and boards or branches may be so arranged that the bees, when they leave the hive, will notice at once the alteration that has been made in its position. This will cause them to mark the new situation and to return to it. The impediments may be reduced on the next day, and removed altogether on the day following, provided that the bees have been flying freely meanwhile. When stocks are being moved from one place to another in the same apiary, and more than two feet at a time, similar precautions must be taken, with this addition—that the old sites must be altered as much as possible in appearance, any bees collecting there being carried back in the evening to their hives. Further, it is to be remarked that moving bees two feet per day can be safely done only on days on which the bees fly, so that they will have marked the position after one move before they be moved again: also, that as a general rule, in winter months, when the bees have been confined to their hives by stress of weather for not less than a few weeks, they may be moved any distance with safety, because they will naturally mark their new position after having been for so long imprisoned.

157. Moving Stocks in Skeps by Road or Rail.—When moving established stocks by car, cart, rail, or steamer, account must be taken of the risks which they will have to run over bad roads and indifferent springs, or at the hands of careless railway shunters and porters; and also of the risks which the public and animals in the neighbourhood will be exposed to should any accident release the bees *en route* and give them "cause of action." For stocks in skeps, an old-fashioned and a useful precaution is to push three or four stout wooden skewers through the skep and combs, from side to side, two or three days before the moving. The skewers, fastened by the bees, act as stays to the combs, and can be withdrawn after the journey. To further reduce the risk of combs breaking away from their attachments, the skep is travelled bottom up. It is first covered with some ventilating material, as directed above **(153),** and is then inverted, placed in a large lidless box, and packed underneath and around with straw. A rope handle is attached to the box, and also a label in a prominent position, and bearing the words—" LIVE BEES, AND HONEY COMB: WITH CARE." An improvement upon these precautions would be—to travel, yourself, with the bees.

158. Moving Stocks in Frame Hives by Road or Rail.—Stocks in frame hives can generally be transported with safety when the following instructions are observed. A hole 4″ × 4″ is cut in the floor board and is covered with perforated zinc. Two lengths of loosely-made straw or hay rope are placed on the floor board at right angles with the frames. Frames with honey and no brood are removed, their places in the hive being filled with frames of old, empty combs, or with empty frames having bands of canvas tacked on from top to bottom, or from end to end, to which the bees may cling. Soft, new combs with brood, if to remain in the hive, are tied in their frames with two broad bands of calico or canvas running under the combs and fastened over the top bars. The dummy is moved up and screwed in position. Instead of the sheet and quilts, a piece of coarse canvas or perforated zinc is laid on the frames and tacked down. Two strong laths are laid across the frame shoulders, and are securely screwed to the hive. The frames, caught thus between the laths above and the straw ropes below, cannot shake about. In the evening, when all the bees are at home, the doors are removed, and the entrance is covered with a piece of perforated zinc securely tacked to the wood. The body box is then screwed to the floor board. The extra frames, roof, doors, etc., travel separately. With large stocks, and in very warm weather, it is advisable to leave in the hive only sufficient bees to cover the brood, and to travel the remainder in a skep as directed for swarms **(153),** hiving them in the usual way **(236)** on arrival at the new locality.

159. Commencing with Driven Bees.—In districts where skeppists follow the barbarous custom of smothering bees at the close of the season, it is generally possible to obtain "condemned lots" at a trifling expense, the owners often being willing to accept a shilling or two for the bees which, otherwise, would be destroyed. When two or more condemned lots can be procured and united on combs of honey, or on frames of comb—if in time to be fed up before cold weather sets in **(315)**—they usually turn out well in the following year. All that is necessary is to examine the stocks for signs of disease, and if they prove healthy, to get the bees away from their combs. This leads on to the operation of "Driving," which, although looked upon by the uninitiated as a wonderful act of legerdemain, is really one of the simplest operations connected with modern bee-keeping.

Fig. 68.
DRIVING BEES.

160. Driving Bees.—A fine day, when bees are flying freely, is to be preferred. The appliances required are—(a) One or two empty skeps, or a Driving box (Fig. 70); (b) Driving irons (Fig. 69); (c) Smoker (Fig. 53, page 67); (d) Bucket (Fig. 68); and (e) a table or chair. Blow a puff or two of smoke into the skep containing the bees, and give them time to run up into the combs, and to feed **(167)**. Carry the stock to some sheltered corner, placing on its stand a box, or an empty skep to decoy any flying bees. Blow some more smoke into the occupied skep, causing the bees to gorge themselves with honey. Place the bucket on a table or chair; lift the skep, invert it, and place it, bottom upwards, in the bucket; set a second skep upon it, "like a cockle shell half open," the skeps

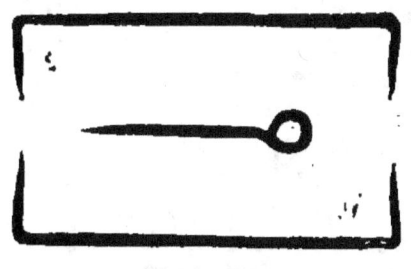

Fig. 69.
DRIVING IRONS.

touching above the ends of two central combs (Fig. 68). At that point push in the skewer (Fig. 69) through the edges of both skeps to hold them together, and stay up the empty skep by the other irons, the points being pushed into the sides of the skeps. These irons are from 15" to 18" long. Two laths, with nails driven through the ends, and a skewer of hard wood, may be made to serve the purpose. With the opening between the skeps in front of you, so that you may observe all that occurs, rap the sides of the lower skep sharply with the palms of your hands, or with two sticks, taking care that while jarring the combs slightly you do not loose or break them down. Carry on the rapping continuously at the rate of about two per second. The bees will speedily run up past the skewer into the upper skep, and if a queen be there, careful watching will discover her passing up. "Close Driving," which is necessary in inclement weather, consists in fastening the skeps together edge to edge, tying a cloth round them, and driving as above; but close driving does not permit one to watch the progress of the operation, nor to see the queen going up. Driving, whether open or close, may usually be completed in about a quarter of an hour. In unfavourable weather, and if there be little honey in the skep, it will be an assistance to sprinkle the combs and bees with warm, thin syrup (Recipe 321) five or ten minutes before driving (181). Colgan's Driving Box (Fig. 70), which was first exhibited at the Armagh Show of 1903, by Mr. William Colgan, supplies the places of skep, irons, and swarm carrier. It is a box, 11" × 11" × 9", with a fast top and a sliding bottom. An iron skewer at the back, moving in two staples, and two pointed iron rods on the sides, working on pivots, hold the box in position upon the skep from which the driving is to take place. The irons, when not in use, fold up on the box. On the lid is a brass handle, and inside at the top is an arrangement for fasten-

Fig. 70.
COLGAN'S DRIVING BOX.

ing pieces of comb or foundation on which the bees may cluster. The front and the sliding bottom are ventilated with perforated zinc. When a stock is to be driven, the box is fixed upon the skep as shown, and, after the operation, the bottom is slipped into its place, and all is ready to be carried away. When all the bees are driven, put them and their skep, or box, back on the old stand in place of the empty skep left there to receive flying bees, which bees should now be shaken out on a board before the entrance to the skep or box containing the driven bees. If two or more driven lots are to be united **(249)** having queens of different values, only the best queen should be allowed to remain. The bees in each lot should be thoroughly dusted with flour from a dredging box or sprayed with thin, scented syrup from an asperser to make them unite peaceably. The Asperser (Fig. 71) is sometimes used for the purpose;

Fig. 71.
ASPERSER.

but spraying with syrup,—a messy, troublesome expedient at the best, is not to be recommended. Since the discovery that ordinary flour will serve the purpose as well as scented syrup, the kitchen dredging box has come into favour as a cheaper, and less troublesome pacifier. When this has been attended to, the two skeps can be brought with the bottoms together, and dumped on the ground, so as to throw those in the upper skep into the lower one. The bees, being then shaken together thoroughly, and having the same scent, will unite peaceably. In the evening, when they have settled down, they can be carried off, or forwarded per rail or post, as described above, and placed upon their new stand, the canvas or zinc being removed. Such lots should be fed up liberally and rapidly. **(315)**. [See also Illus. p. 99, and "Automatic transfer from Skep to Modern Hive." **(254)**.]

161. Study the Subject.—To begin intelligently, study the subject thoroughly. Make yourself familiar with the nature and habits of bees and with the most improved methods of management. If you have an experienced bee-keeper in your neighbourhood, or among your friends, gather from him all the information that he can supply, and ask him to allow you to witness his manipulations from time to time. But, when you have studied a Bee Guide and have seen some of the operations connected with bee-keeping, do not suppose that you can afford to proceed with the industry without keeping yourself in touch with the approved literature of apiculture. In order to take advantage of the latest discoveries, and of the experiences of the foremost bee-keepers of the day; in order to combine with the interests of the pursuit the profits which it is capable of providing, you should subscribe for a reliable publication and thus acquaint yourself with what is being accomplished elsewhere, with the developments which are taking place, and with the views of the most capable apiculturists upon the innumerable questions which, although outside the purview of a guide book, present themselves in actual experience every day. The *Irish Bee Journal**—the Official Organ of the Irish Bee-keepers' Association and its affiliated Associations, and the *Beekeepers' Gazette**—the Official Organ of a large number of the most progressive County and District Beekeepers' Associations in Great Britain, are edited by the author of this Guide, and supply information upon every subject connected with beekeeping. Queries addressed to the Editor are replied to either direct per post or telegraph, or in the "Expert Advice" columns of the Journal and Gazette. (See Note, page 209)

BY MOTOR TO THE HEATHER.

---

* *Irish Bee Journal, 3d. monthly, 4s. per annum, post free. Beekeepers' Gazette, 3d. monthly, 4s. per annum, post free. From the Office* BEE PUBLICATIONS, *Lough Rynn, R.S.O., Co. Leitrim, and from all newsagents. Wholesale Agents.—W. H. Smith & Son, Strand House, Strand, London, W.C.2; Eason & Son, Ltd., Dublin and Belfast.*

CHAPTER XVII.

SUBDUING AND HANDLING BEES.

162. Tranquillising Influence of Smoke.—When all due homage has been paid to those great scientists whose discoveries and inventions have led up to the present highly developed condition of apiculture, there yet remains a tribute of praise and gratitude to the man (whoever he may have been) who first disclosed a plan by which the bee may be subdued and reduced to a temper so amiable as to be amenable to handling without showing fight. For, it can hardly be doubted that all the knowledge of bee instincts which has been attained

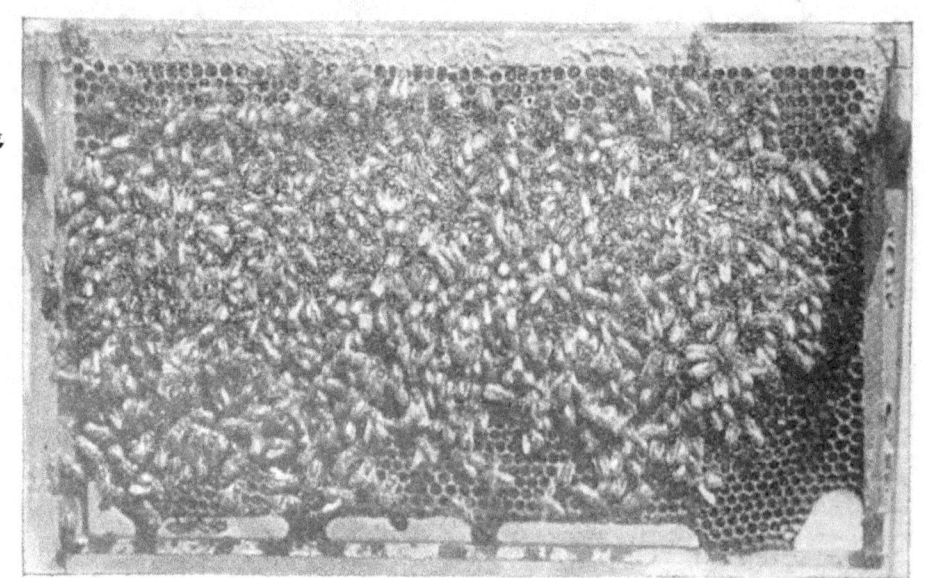

Photo from life]   Fig. 73.   *[by J. G Digges*
SUBDUED BEES ON EMPTY COMB.

and all the improvements in bee appliances which have been effected in modern days, could not avail to bring the industry to its present stage of progress had not some method been devised for breathing a peaceful calm over the occupants of a hive under manipulation. All honour to the man who first discovered the tranquilising influence of smoke!

163. Unprovoked Stinging Exceptional.—It is an utter fallacy which suggests that the main object, or one of the main objects of a bee's life, and her greatest happiness, is to drive her sting into human flesh. The sting is her natural protection—a weapon not properly of offence, but of defence. It is the exception, and not the rule, for the sting to be used in a manner unprovoked. Instinct teaches the bee to employ her weapon sparingly, because the fastening of the barbs in the object stung often obliges the bee to retire mutilated from the encounter. (39).

164. Fearless Defence of the Home.—But bees, it must be admitted, sometimes conceive extravagant notions of danger, and, without any cause apparent to us, will attack with fury any other living thing in their neighbourhood. In such circumstances discretion will often prove "the better part of valour." Force is no remedy. Attack them with your umbrella; a hay fork; a locomotive; a pom pom; they will beat you. Bring up the British Army, horse, foot, and dragoons; the bees will win the day. For behind their assault are their queen; their brood; their home; and in defence of these they are utterly oblivious of danger and indifferent to death. For their fearless anger when aroused, bees have been employed in warfare. There are cases on record in which whole regiments have been routed by the letting loose of bees. In Thuringia (1525) a furious mob, which had stood out against tremendous odds, was instantly put to flight by having hives of bees thrown among them. To the uninitiated there is something terrifying in the vicious buzzing of bees when they have their abdomens curved for the thrust, and the very air around them seems charged with venom. You cannot oppose your courage to theirs, for they are not amenable to the laws of civilised warfare, and they will fight with irresistible bravery, and will die a thousand deaths, if need be, in defence of their homes.

165. What Constitutes "a Master of Bees."—Therefore it is necessary, in order to manage bees, whether on the old principles or the new, that one should know how to stay "the beginning of strife," to subdue them to his will, and to bring them completely under control. Firmness, without aggression; gentleness, without fear; and a knowledge of their habits, tastes and fancies, are all that are required to constitute a master of bees. With such qualifications one can do with them as one pleases; can revolutionise their kingdom; depose their queen; regulate their enterprise; intercept their swarms; order the manner of their industry; deprive them of

94 THE PRACTICAL BEE GUIDE.

their stores; and, without provoking their anger, turn them again to peaceful labour. It is not a charm that may be worked by a privileged few. It is the application of a knowledge to which all may readily attain.

166. Swarming Bees Harmless.—It is well known that bees of a swarm are usually as harmless as butterflies (19). They may be gently lifted in the hand, and dropped, bunch after bunch, without so much as an angry buzz from them. A gentleman carried through the noise and bustle of city streets a swarm that had settled on his head. He walked with them into his office, and secured them in a box. They made no attempt to sting him. There must be some reason for this. Visitors to a Bee Tent look with amazement upon the lecturer driving bees from skep to skep; picking them off the combs; remaining unmoved with bees crawling upon his neck, or hanging to his eyebrows. The onlooker cannot understand it. Yet it is easily explained.

Photo from life]　　　Fig. 73.　　　[by J. G. Digges.
SUBDUED BEES WITH CAPPED BROOD.

167. Full of Sweets—Empty of Bitterness.—Before issuing in a swarm, it is the habit of bees to fill their honey sacs from the stores, instinct teaching them to carry from the home, which they are about to abandon, sufficient food with which to secrete wax for new combs, and to support themselves in the interval (18). In that condition they are most peaceably disposed, and will not sting except under violent provocation. If they can, at other times, by any means, be brought into a similar condition the same results will follow. It has been found that any sudden, mysterious alarm communicated to all the colony in a hive will drive them to the honey cells for food. A puff or two of smoke blown in at the entrance (126), or a carbolic cloth (127) laid on the frames, has the necessary effect, and a peep under the quilts will then discover the bees with their heads in the cells, drinking deeply. A delay of a minute or two, and the whole colony will be found subdued. With gentle handling the frames may be taken out; examined; hung upon a stand (Figs. 72, 73, and 77) and returned to the hive, not a bee taking wing, nor any attempt being made to show resentment. In this manner the fiercest colony may generally be subdued and handled with safety. With Maeterlinck this inoffensiveness is the result of happiness; with Simmins, of homelessness; with Cheshire, of terror; with Langstroth, of a physiological fact—

"When a bee prepares to sting, she usually curves her abdomen so that she can drive in her sting perpendicularly. To withdraw it, she turns around the wound. This probably rolls up its barbs, so that it comes out more readily. If it had been driven obliquely instead of perpendicularly, as sometimes happens, she could never have extracted it by turning around the wound. When her stomach is empty, a bee can curve her abdomen easily to sting. If her honey sac is full, the rings of the abdomen are distended, and she finds more difficulty in taking the proper position for stinging."—*Langstroth.*

168. A Firm and Gentle Hand Necessary.—It is known, also, that bees resent roughness; have a deep-rooted objection to jarring of their combs; fly into a passion if any of their number be crushed in the hive; will not tamely submit to being rubbed the wrong way; are provoked to violence when one sting has been inflicted, by the mere smell of it; and will often attack a hand for no other reason than that it has been suddenly and quickly moved adjacent to them. Sometimes bees noted for their peaceable disposition, will be found in an angry humour, some unaccountable influence having disturbed their wonted calm. Therefore bees should always be handled with the utmost deliberation and care. A firm and gentle hand is necessary. There must be no jarring of the combs, no swiping of the handkerchief at a threatening bee. Coolness gained by experience, together with the precautions already, and yet to be described, will usually render the manipulation of bees as safe as the driving of a flock of geese. Should bees at any time show marked signs of vindictiveness, it is better, instead of attempting to fight them, to withdraw in a manner as quiet and dignified as possible (179). They will probably be found, next day, in their normal condition of amiability. To start in and fight them, may render them unmanageable for the remainder of the season, and will certainly lead to a precipitous flight.

"*Effect of Stings.*—A writer in a contemporary wants to persuade us that formic acid is not volatile. He ought to observe its effect on dogs, and even on the slow-footed donkey. A venerable angler, coming too close to a concealed apiary last summer, was seen, inspired with marvellous energy, laying about him with his fishing-rod—when you come to think of it, an absurd weapon for the occasion. After that he took a five-bar gate with the agility of a youngster. The fishing-rod was propping asters in September! We once saw a boy scale a seven-foot wall, glazed, and drop into the street on the other side, as it seemed from simple enthusiasm. The gate stood open all the time within a jump of him. He had been trying to scrape honey out of a hive entrance with a three-pronged fork. Formic acid not volatile. Stuff and nonsense."—*Irish Bee Journal.*

169. Protection for Beginners.—Beginners, and all who have not yet gained confidence from experience, will do well to remember that after bees in a hive have been thoroughly subdued, there may be an occasional bee returning from the fields, or dodging about the hive, which has not yet been influenced for good; and that, until it can safely be dispensed with, a veil will prove a most useful protection for the head and neck against the attacks of the "free lances" of the colony **(128-130)**. Procure a hat with a broad brim (Fig 54); draw the veil over it until the elastic grips the lower part of the crown; settle the veil over the shoulders; button the coat, to keep all secure; and see that the veil is at least the length of a bee-sting apart from the face, ears, and neck. Next provide against the possibility of a bee crawling up your legs, and fasten your coat cuffs to protect your arms, for pressure of the clothes will, certainly, cause a bee there to sting. If you find it necessary to do so, don a pair of bee gloves, to protect your hands and wrists **(131)**. Thick woollen gloves, though safe, are not desirable, because it is difficult to manipulate with them, **(132)** and, although you may not be hurt, bees will often sting them, mutilating themselves in the operation **(39)**. Apart from the fact that one can never afford needlessly to sacrifice bees, it is not humane, nor in accordance with the principles of modern bee-keeping, to provoke them to leave their stings in one's apparel. Burkitt bee gloves offer little hindrance to manipulations, and are seldom attacked **(132)**. It should, however, be the aim of every beginner to dispense with the use of gloves as soon as possible

A suitable attire for ladies (fig. 74) has been illustrated and described in the *Irish Bee Journal* as follows:—

Fig. 74.
BEE DRESS FOR LADIES.

"A white smock, made of washing material. It buttons tightly round the neck, over the veil, and down the back, being secured by a belt round the waist. The smock keeps the dress free from honey, vaseline, etc., and can be constantly washed, which is an advantage if foul brood has to be dealt with. A pair of Burkitt gloves are drawn over the hands and the ends of the sleeves, and with a wire veil, the stings of the bees are defied. Of the Burkitt gloves I cannot speak too highly No sting seems able to penetrate them, they are not clumsy to work in; and they give much confidence to nervous manipulators."

170. Treatment of Stings.—If the hand be stung, and the sting be left in the flesh, the sting should be withdrawn immediately, not squeezed, but drawn out with the nail or a knife-edge; because the reflex action will continue for some time to inject poison into the wound if the sting be not removed (33). If a drop of ammonia be at once applied to the wound, the pain and swelling may occasionally be reduced; and if the spot be touched with the carbolic feather (176), the bees will not be excited to further attack by the smell of the sting poison (168). External applications, however, cannot be relied upon to neutralise the injected poison. If, when the sting has been removed, the part stung be not rubbed, but pinched with the finger and thumb until, on loosing it, the pain does not return, little trouble will be experienced. It generally follows that, when one has been frequently stung, one becomes safe from pain and swelling as results of stings. Further, there is much testimony to the fact that such ailments as rheumatism are alleviated and even cured by a sufficient application of the sting of the bee; so that the pain of the sting is not without its compensation.

"I am a firm believer in the efficacy of stings as a cure for rheumatism. Shortly after my recovery from rheumatic fever, a lady presented me with an entire apiary, and in the transfer of the stocks I got a 'murthering' of stings, and, though I had been subject to rheumatism for years previously, I never, since that stinging, felt a twinge of it."—T. B. O'Bryen, in the *Irish Bee Journal*.

Photo by]   USING THE SMOKER.   [J. G. Digges.

CHAPTER XVIII.
MANIPULATING.

MISS W. SEADON (Aged 7 Years) DRIVING BEES.

171. Appliances Required.—Before opening a hive for manipulation, be careful to have at hand everything that you may require. A smoker **(126)**, a carbolic cloth **(127)**, a small table that can be carried from hive to hive, a comb stand **(172)** to hold frames of foundation and frames removed from the hive, a comb box **(173)**, a dinner knife, a wing or soft brush, a pot of vaseline or petroleum jelly **(174)** are all useful articles.

Fig. 75.
COMB STAND.

172. The Comb Stand (Fig. 75), is intended to hold frames when a hive is being manipulated **(185)**. It is often necessary temporarily to remove one or more frames from a hive when operations are in progress; and it is always useful, when working at hives, to have spare combs at hand in convenient position. The stand shown holds three frames on each side. It is 2' 6" high, and the carriers, fastened on the legs, are 14½" apart. The stand can be carried about the apiary, and set down where required, without danger of breaking the combs or of injuring the bees that may be upon them. Its usefulness may be further observed on referring to the illustration facing page 80, and to figures 72, 73, 77 and 97.

Fig. 76.—COMB BOX.

173. Comb Box.—When combs are being transferred, removed for extraction, or carried about the apiary, it is advisable, in order to minimise the risk of robbing **(308)** to have a comb box in which they can be placed. The comb box illustrated (Fig. 76), is internally 17½" long × 9½" deep × 9" wide. Two carriers are nailed at the ends inside, 1" from the top, to take the shoulders of the frames. A handle is fixed on the lid, and a cone escape **(273)** permits the exit of any bees that may have been shut inside.

174. Vaseline, or Petroleum Jelly, is applied to the shoulders of frames, to the carriers on which they rest, and to the bottoms of crates, etc., to prevent the propolising of them by the bees **(266)**. It is so desirable to have all hive fittings easy of removal, without jarring, that the application of vaseline or petroleum jelly should never be omitted by the bee-keeper who desires to perform his manipulations without needlessly provoking his bees **(168)**. The material, which is inexpensive, may be applied with a small paste brush.

175. Preparing the Smoker.—The smoker **(126)** should be in good order, and the fuel prepared beforehand, for it is most disconcerting to have the smoker give out when operations are in progress. Almost any dry fuel that will burn may be used—dry, rotten wood, rag, or brown paper. Put a couple of quarts of hot water into a bowl; dissolve in it, say, one or two ounces of saltpetre; soak a quantity of brown paper in the liquid, and when dry, cut it in strips about four inches wide. Roll one of the strips loosely; light one end, and put it into the smoker, lighted end down. Small rolls of dry brown paper may be added from time to time as the fuel in the smoker becomes exhausted.

176. Preparing the Carbolic Cloth (127).—Procure from any chemist a bottle with an asperser cork. In this make a solution of 1 part Calvert's No. 5 Carbolic Acid to 10 parts water. Take a piece of ticking, calico, or linen, say 24" × 18", which, in some operations, may be more conveniently used if prepared like a flag (127). Shake the bottle and thoroughly damp the cloth with the solution. Sprinkle a little also on the feather. Put cloth and feather into a tight-fitting tin box that they may retain the odour.

177. Opening the Hive.—Go, now, to the hive which you want to examine. Blow one or two puffs of smoke through the entrance, into the hive, remembering that your object is not to half smother the bees, but just to send them to dinner. Rap smartly with your knuckles on the sides and roof; set down your smoker, nozzle up, so that it may draw like a chimney; take out your carbolic cloth and feather, and wait for a minute before proceeding further. Then place the feather half its length into the entrance, to put a stop on the bees there; and remove the roof and quilts, leaving only the sheet on the frames or super. By this time the bees will have gorged themselves into good humour. Your position will now depend upon the arrangement of the frames in the hive. If they hang at right angles to the front, stand at the side; if they hang parallel to the front, take your position at the back. Hold

DRAWING ON CARBOLIC CLOTH.

the carbolic cloth by the lath, if one has been inserted (127), or by two corners, and let it hang down outside the hive at the side opposite to you. Pick up the corners of the sheet, and slowly draw it back upon itself towards you, so bringing the carbolic cloth over the frames as illustrated. Not one bee will get out if you do this carefully. Instantly the bees will begin to make music—a peaceful symphony which may encourage you. Remove the cloth, or roll it back off two or three frames—you will find the bees with their heads in the cells, or moving about in a bewildered fashion, gentle as lambs, and disposed to treat you with every courtesy if only you reciprocate their gentility.

178. Manipulating Wicked Stocks.—In the case of a wicked stock of Natives **(46)**, Cyprians **(49)** or Syrians **(50)** the subduing may require to be of a more thorough-going nature. Give three or four puffs of smoke at the entrance, and close the doors; with your shut fists drum on the roof for half a minute; open the doors, and give more smoke; and drum again for half a minute. After three or four minutes draw on the carbolic cloth, and the bees will probably be found perfectly subdued. Keep the smoker at hand, to drive them back if they should show a desire to boil over.

179. Forcing the Pace.—In the event of a usually quiet stock proving unruly, as will occasionally occur **(169)**, suspend operations at once, and withdraw. Give them time to calm down, and try them again on the next day. Bees, like mortals, sometimes "get their dander up," and probably with better reason. They may have been fighting robber bees **(310)**; they may scent rain in the distance: they may have been provoked by some interfering man or beast—you cannot always tell. But it will be better to let them "sleep upon it" than, by forcing the pace and persevering in your manipulation, to run the risk of turning them into demons for the rest of the season.

180. Smoking Overdone.—It must, however, be said, that with our Native bees **(46)**, Italians **(47)** or Carniolans **(48)** elaborate preliminaries to manipulation are not often necessary. When you have gained experience, and have learned how to do it, you will frequently find yourself able to open and manipulate a hive without the aid of smoke or carbolic, though these should always be at hand in case of any emergency. In the honey season, thoroughly smoking a colony puts a stop to the gathering of nectar, probably for the rest of the day. The honey that has been gorged has to be disgorged into the cells when you have finished operations. If you stand aside and observe, you will find that nothing like the same energy is displayed at the entrance, and if you weigh the hive next morning, it will be seen that the average increase has been suspended; and that a loss of from 5 lb. to 10 lb. of honey has been incurred. Therefore, smoking should never be overdone; and for simple operations, such as putting on or taking off a super, it is seldom necessary at all **(266)**, nor, indeed, unless the brood nest is to be disturbed, or the hive manipulated at unsuitable hours.

181. No Food—No Subjugation.—It must also be added that the use of smoke for quieting bees presupposes the existence of

MANIPULATING. 103

food in the hive. With the best intentions possible, bees cannot gorge themselves with smoke. If, therefore, there be no food in the hive upon which they can feed liberally, warm syrup may be given. The carbolic cloth can be drawn over the frames as described above, and in a few seconds the syrup may be lightly sprinkled between the combs, the smoker being applied subsequently.

182. Examining the Combs: Finding the Queen.—Having subdued the bees, as described, you may proceed to examine the combs. You must remember that the space between the frame ends and the hive is not more than ⅜", perhaps only ¼", perhaps less; and that if you draw out a frame carelessly you run the risk of crushing bees, and even of killing the queen, if she happens to be on one of the ends of the frame (83). Draw back the dummy (94) as far as it will go. Draw back the frames together from the centre frame, *i.e.*—if there are nine frames in the hive draw back Nos. 1 to 4 together, and very

Photo from life]   Fig. 77.   *[by J. G. Digges.*

SUBDUED BEES, WITH CAPPED WORKER AND DRONE BROOD.

slowly, so that the bees on the ends may have time to get out of the way of danger. If you have, up to this, kept the carbolic cloth on the frames, and have kept daylight out of the hive, the queen will probably be found on the centre frame; but if you have flooded the brood-nest with light, she will have made off to the front or the back frames. Take the centre frame

104   THE PRACTICAL BEE GUIDE.

by the shoulders, in your fingers, and lift it vertically over the hive. Turn your back to the strongest light, and examine the side of the frame next to you. If the queen is there you will recognise her by her length, and shape, and colour (4). She is longer than the worker bee; thin for her length; with her abdomen pointed; and somewhat darker in tint than the others. If young, she will show her activity by "dodging" from one side to the other of the frame. If she be on that frame, it will be well to return it to the hive, lest she take wing, and give trouble; in which case you must remain perfectly still and await her return. (See also **185b**, page 107.)

183. The Combs Described.—Your frame, upon examination in the summer season, will probably be found to contain honey, capped and uncapped, capped brood, uncapped larvæ, eggs, some empty cells, and perhaps a little pollen (Fig. 14). Bees store their honey over their brood. The cells, therefore, next the top bar,—cells capped with light coloured wax, contain ripe honey. Next to them will be found unripe honey, not yet capped. Lower down on the comb, cells with a dark capping

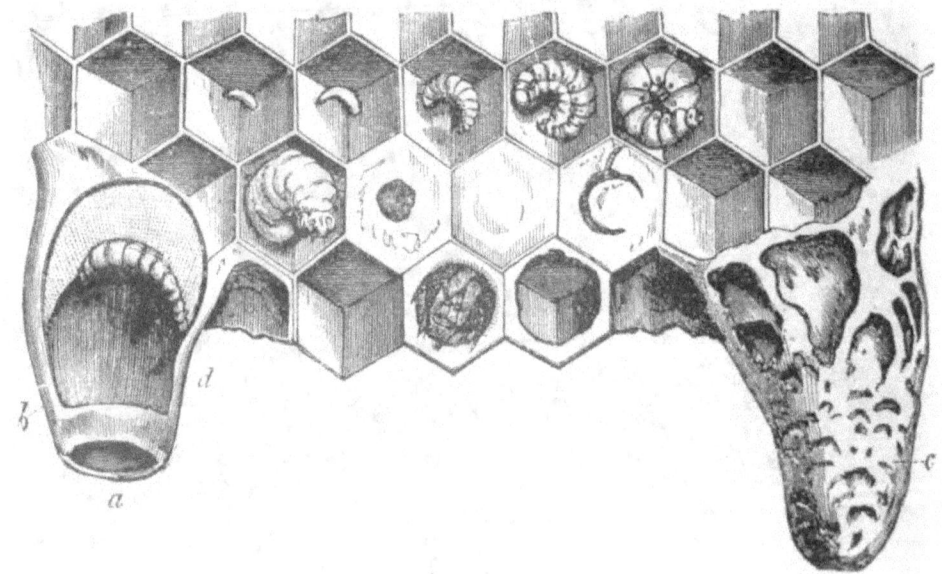

Fig. 78.
COMB AND QUEEN CELLS.
(Magnified twice.)

a, Queen cell, cut to expose "Royal Jelly" and Grub at upper end; *b*, Thickness of cell; *c*, Dimpling outside cell; *d*, Spot where bowel contents and exuvium are placed.

of wax and pollen contain hatching brood, the wax being mixed with pollen to render the cappings porous. Some of these capped cells stand out from the comb beyond the others (Fig. 77), and have a larger diameter; they contain drone brood (Fig

14, D). Perhaps one or two cone-shaped cells appear, hanging down; these are queen cells (Fig. 78 and Fig. 14, page 36). Close at hand open cells will show the larvæ, pearly white, in various stages of development (Fig. 78). And others have eggs (189), like little bits of blue-white thread, on the bases of the cells. Some of the eggs stand out at right angles from the comb; these are one day old: others bend over towards the base; these are two days old: others lie upon the bases of the cells; these are three days old, and are just about to produce larvæ. Other cells, capped and uncapped, contain pollen, or "bee bread," of various hues. As bees store pollen near the entrance, brood in the middle, and honey at the back, if your frames run from front to back, you may find pollen, brood, and honey in the same comb; while if the frames run from side to side of the hive, pollen will probably be found in

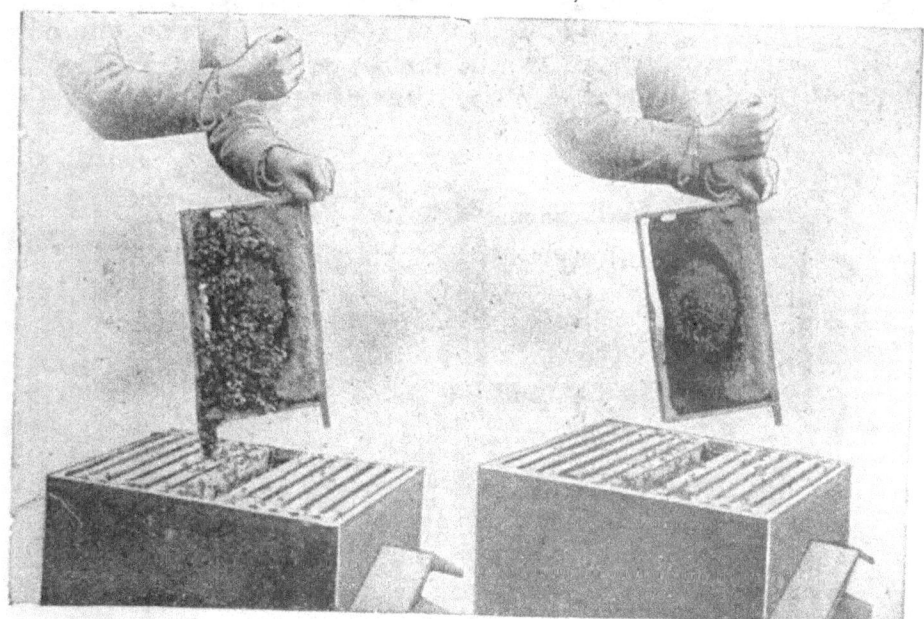

Photo from life]   *[by J. G. Digges.*
Fig. 79.
"THUMPING" BEES OFF A COMB.

the combs next the entrance, honey in those at the back, and brood, with honey higher up, on the other frames.

184. Removing Bees from Combs.—To get bees off a comb, it is sometimes advised that they be jerked off or brushed off. They may be jerked off if one keeps on jerking long enough, but the second jerk often puts bees on the wing, and leads to mischief. They may be brushed off with a goose wing, or a strong feather, but, though bees will often submit meekly to a brush that meets them in the face, if it take them the other way—(as it must, since they are on the frame heads and tails)—they are apt to rise to the occasion in a manner that does not always appear to the operator to be quite justified by the circumstances. There is another plan. Hold the frame firmly by one of the shoulders in your left hand, keeping it a few inches over the hive (Fig. 79). Now, with your right fist give a sharp thump on your left hand. To the bees it will be like an earthquake, and a thing irresistible. They will drop, to a bee, and scamper down among the other frames as if the end of their world had come. Of course, if you let the frame drop, the "earthquake" may possibly astonish yourself. Keep a good grip, and never employ jerking or thumping with a frame that has upon it a queen cell, or much unsealed honey.

185. Turning Combs.—To examine the comb on the other side, you must not turn it up as you would a slate. A comb so used, especially if new, or heavy with honey, may drop from the frame, or sag, and break; or may drop honey out of the cells. Therefore, whether the combs to be handled be old or new, wired or not wired **(262)**, make it your rule to turn them in this manner—

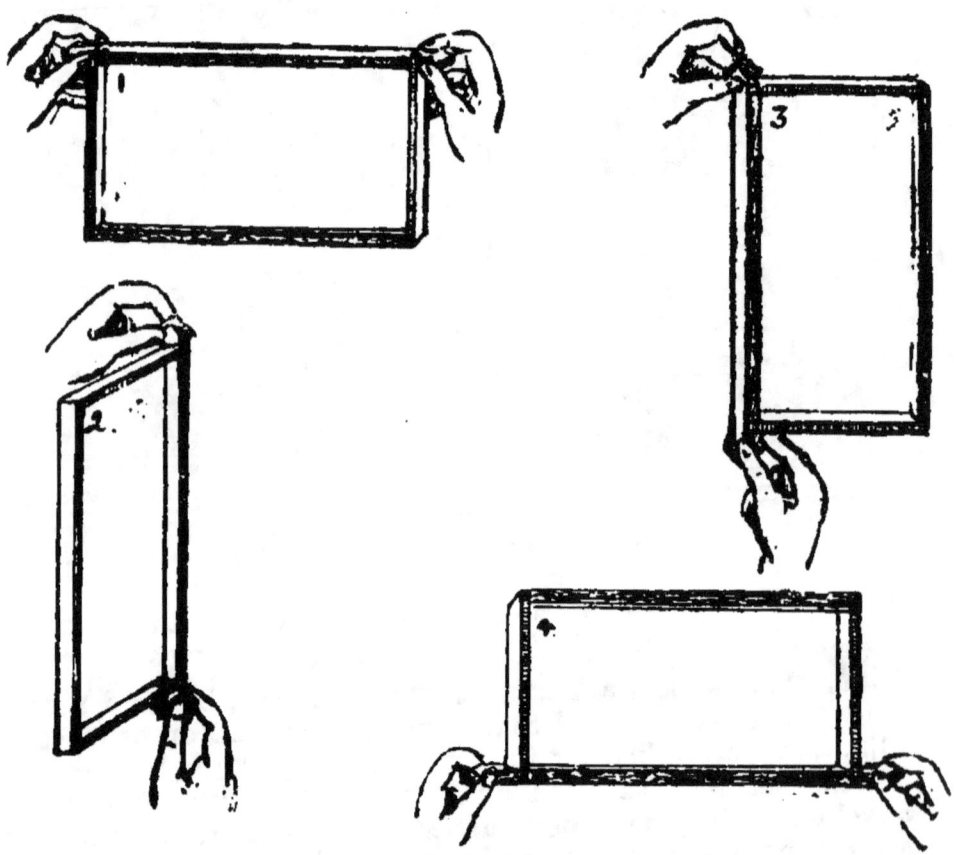

Fig. 80.
TURNING A FRAME.

Hold the frame by the shoulders (Fig. 80, 1): raise your left hand, bringing the frame into the position shown (2): give the frame a half turn, like a swinging door, bringing the off side next yourself (3): then lower the left hand into the position shown (4). Reverse these movements to bring the frame back to position 1, Fig. 80. If the queen is not upon it, you may now hang it on the comb stand **(172)** and proceed to examine the other combs. In all these operations, act with deliberation; move your hands slowly; avoid pinching bees with your fingers or crushing them in the hive; and do not jar the frames when putting them back, nor kick the hive legs with your feet. If the bees show signs of getting from under control, draw the carbolic cloth over the frames again, or blow a puff of smoke along the frame tops. When you have finished your inspection, replace the frames, shoulder to shoulder; draw up the dummy; put on the sheet and quilts and roof; and, if the bees have not already thrown it out of the entrance, remove the carbolised feather.

185b. Searching for the Queen.—In the fascinating game of hide and seek with the queen **(182)**, there are yet other rules to be observed. For example, a comb on which the queen may be, when lifted out for inspection, should be held over the hive, so that in the event of her falling off she may drop safely into her home. Combs containing brood should not be hung upon the comb stand **(172)** for any length of time in chilly weather lest the brood be injured **(338)**. Combs containing much unsealed honey should not have their adhering bees removed by jerking or thumping **(184)**, for this would throw out the honey and do damage to the bees **(36)**. Stand with your back to the sun, and as each frame is withdrawn scan the exposed faces of the combs in the hive, for her majesty may be scampering there towards the darker side. Part any cluster of bees on the frame in your hands, the queen may be hidden there. If still she baffles you, and if it be necessary to find her there and then, either of the following methods may be tried:— Place a hiving board in position **(233)**, a piece of excluder zinc **(109)** on the hive entrance, and a dummy **(93)** inside near the hive front. Lift out the frames one by one and thump, or brush, every bee on to the hiving board, moving back the dummy and returning the frames in front of it as they are cleared; when the excluder stops the queen entering you will have her. Or, procure a super box **(108)** or any bottomless box about the size of the hive, invert it, and brush every bee into it from frames, hive, and floorboard; set up the hive again with the frames in position; place an excluder on the box, set the latter on the frames, with the excluder between, thump on the lid to throw the bees down, then raise the lid and spread a carbolic cloth **(127)** over the box; the bees will quickly run down to the frames, except the queen and drones, which will be trapped above the excluder.

CHAPTER XIX.

BREEDING.

186. Breeding begins.—Breeding in the hive generally begins towards the end of January **(8)**, perhaps somewhat earlier in a mild season. Moving quite slowly upon the centre combs, the queen, examining the cells and inserting her abdomen, deposits her eggs upon the bases—one egg in each cell, confining herself at first to a small area, and increasing the areas as the season advances; passing from comb to comb, and returning to the cells according as they become vacant through the hatching of the young bees. If, on account of the smallness of the cluster, or the condition of the queen, more than one egg be laid in each cell, the workers will generally remove the superfluous eggs.

187. "Congestion" to be guarded against.—A comb completely filling a standard frame contains 104 superficial square inches on each side, or 208 superficial square inches in all. Worker cells measure 27 to 29 to the square inch **(66)**. Taking them as 28 to the square inch, we have 5,824 worker cells in the comb of a standard frame. Given sufficient room and favourable circumstances, a queen in her prime, laying at the rate of 3,000 eggs per day **(4)**, or 90,000 eggs per month, might occupy with eggs more than half of such a comb per day; and nearly 16 complete combs per month. But the eggs deposited on the first day of the period will hatch out, and the cells become vacant, on the 22nd day **(204)**. Assuming that the queen returns always to the cells as soon as they become vacant, she might, at the rate of 3,000 eggs per day, have filled 11½ complete combs in the twenty-two days before she returns to the former cells. From which calculation it will be seen that, in the height of the season, and with a queen in her prime, the increase of the colony will be very rapid; and that, in a hive containing no more than 10 or 11 standard frames, and with from 30,000 to 50,000 bees depositing honey in the cells, the queen may very quickly find herself hampered for room, and that "congestion" supervene which produces the "swarming fever" **(216)**. It follows that, where the largest possible harvest of honey is desired in preference to an increase in the number of colonies by natural swarming, the bee-keeper must

BREEDING.

so arrange that the queen shall always have more room than she actually requires for the depositing of her eggs, and that the bees shall have, at the same time, sufficient room for the storing of honey. This is what is referred to by the frequent advice to "give room in advance of requirements," so that congestion shall not provoke swarming, and thus disorganise, in the middle of the honey flow, the work of the colony. **(216).**

188. Drone-breeding Queens.—Until the approach of the swarming season, the queen lays only impregnated, *i.e.*, worker eggs; after which drone cells are prepared, and in them she deposits unimpregnated, *i.e.*, drone eggs. A queen in her fourth or fifth year will sometimes, however, become a "drone-breeder": the supply of fertilising material in the spermatheca **(45)** having become exhausted, she is no longer able to fertilise her eggs; and, though she may continue to lay in both worker and drone cells, the produce from both will be drones only—dwarf drones, if reared in the cells intended for worker larvæ. Such a queen should be supplanted at once; in fact, after her second year, a queen ceases to be profitable, and her place should be taken by a young, fertile queen **(281-2)**. A hive which shows too large a proportion of drone brood should be re-queened without delay.

189. Age of Larvæ.—From Dr. E. F. Phillips, in *Gleanings*, we have the following data for judging the ages of larvæ. Just hatched, a straight line from head to tail is ⅛th the diameter of the cell; one day old, ⅓rd, the form semi-circular; two days, head touches tail in a circle nearly ½ the cell diameter; three days, it occupies ⅔th; four days, it fills the entire diameter of the cell. To be able to tell the age of egg **(183)** and larva is very desirable, especially when arrangements are being made for queen rearing. **(293).**

Fig. 81.
EGGS AND BROOD.
a, Eggs, natural size; *b*, Eggs magnified; *c*, Larva, natural size; *d*, Nymph, natural size.

190. Worker Brood.—During the first three days the germ feeds upon the substance of the egg; and, hatching on the fourth day into a small white grub, it is supplied by the nurses with a food elaborated for the purpose in the stomachs of the nurses. After about three or four days more, a mixture of semi-digested honey and pollen is added to the food. On the ninth day from the laying of the egg, in the case of worker brood, the cell, well supplied with food, is sealed with a porous capping consisting of a mixture of wax and pollen; the larva (Fig. 81, *c*) spins a cocoon, casts off one skin after another, and becomes a pupa or nymph (Fig. 81, d), gradually becoming transformed into a perfect bee; in which condition, on the twenty-second day, she bites the capping and gains her liberty. The cell is at once cleaned and prepared for the reception of another egg. Twenty-four hours later the young bee begins her life-work by acting as a nurse to the larvæ in their cells; a few more days elapse before she flies from the hive; and, about thirty-six days after the laying of the egg, and fourteen days after her exit from the cell, she begins the work of foraging **(204)**, which work, arduous as it is, will exhaust her energies and bring about her death in five or six weeks of summer.

191. New Combs for Breeding.—The adhering to the cell walls of the cocoons spun by the larvæ **(190)** tends to reduce the size of the cells, and eventually to render them unsuitable for the rearing of vigorous bees **(73)**. It is said that combs quite twenty years old have been found capable of producing bees as large and as strong as those reared in new combs; but it is not desirable to retain combs so long for breeding purposes, and modern bee-keeping favours frequent renewal of combs in the brood nest, for the reason indicated, and also upon principles of hygiene. Three or four frames of foundation given to a colony every year, thus renewing the combs every third year, fulfills a rule that is well worthy of general observance.

192. Stimulating in Spring.—For the rearing of larvæ, pollen **(74)**, liquid food, and a temperature of from 80° to 90° are required. To assist the nurse bees, and to encourage the queen to increase the circles of her brood, the bee-keeper begins "stimulative feeding" **(313)** early in the spring, and also adds fresh, warm wraps to preserve the heat of the hive. A cake of flour candy is given over the cluster **(324)**. Later on, once or twice a week, the sheet is quietly rolled off the tops of one or two frames, and the cappings of some of the honey cells are scratched, or bruised, to entice the bees, and to cause them to use the honey. This can be done without removing the frames or disturbing the bees. If the carbolic feather **(176)** be passed between the combs, the bees will move down before it, and, with the flat of a knife, cappings can be bruised so as to expose the honey, which will stimulate the queen and bees to fresh efforts. A feeder **(119)** is placed on the frames directly over the brood nest, and each evening, when the bees have ceased flying, a small quantity of thin syrup—no more than the bees will take down during the night—is given warm, the doors of the hive being closed to about half-inch space to prevent robbing **(310)**. In some districts natural pollen is so plentiful early in the spring that bees will not use artificial pollen. But if they cannot procure the former, the latter must be given **(320)**. Pea flour makes an excellent substitute for natural pollen. It may be dredged into the vacant cells of an outside frame and placed beside the brood combs **(183)**; it may be dropped into the blossoms of crocus or other spring flowers adjacent to the hives; or it may be exposed in the apiary, in a small box, protected from rain, and upon it may be laid a few straws on which the bees may alight. The effect of a good supply of pollen in the spring is often very remarkable. Forming, as it does, a very necessary ingredient of bee food, it stimulates brood rearing and adds an air of busy industry to the whole apiary, with results very desirable to the bee-keeper who wishes to have his colonies built up to full strength before the opening of the honey flow **(255)**.

"An Experiment.—I am the happy owner of a small garden. In it grow many plants and shrubs, and some fruit trees. Among them are a companion pair of pretty daphne shrubs. It is natural to them to come forth in full bloom at the end of February, while there is not yet a leaf to shelter their bare branches. Just at that time we got a few warm sunny days, and forth from every hive came myriads of delighted workers. The air was thick with them, and a few condescended to test the nectaries of the daphne flowers; but the aerial dance of the others was too entrancing to permit their following the lowly example set them. I fancied that this was a good opportunity for the distribution of artificial pollen, and procured a supply of flour, which I dredged over the shrubs. Immediately the shrubs became two living bouquets. Never did I witness such a desire for hard work; and in quite a little time bees were returning to their hives with pellets of pollen as soft as floss. I repeated the experiment twice a day while the warm weather lasted, and I have no doubt but this aided to bring my stocks into the congested condition they now are in, the majority calling visibly for supers at the end of April."—M J. O'Doherty, in the *Irish Bee Journal*.

Under such treatment breeding will proceed apace. And, to still further hasten the growth of the colonies, the operation of "spreading the brood" is resorted to.

193. Spreading the Brood.—Spreading the brood consists in enlarging the brood nest by the insertion, in the centre of it, of frames of drawn out comb, or of comb foundation. This is an operation which should not be attempted by inexperienced bee-keepers. It must not be recklessly performed, nor without due regard to the strength of the colony; because, if the brood nest be enlarged beyond the covering capacity of the bees, brood will be chilled, and much mischief may ensue **(338)**. But, where wisely and carefully carried out, the effect of spreading the brood is to quickly increase the strength of the colony. For, the queen, finding vacant cells in the centre of the brood nest, will deposit eggs in them at a period of the year when she would not be likely to travel to the colder, outer combs for the purpose. As a general rule brood spreading may be considered safe when, the weather being warm and the nights no longer chilly, the space between the outside comb and the dummy (93) is found to be occupied by bees. In this case, if the outside comb be suitable for brood rearing, *i.e.*, a straight, well-built comb, not overstocked with pollen, and preferably containing some honey, it may be used for the purpose. The carbolic cloth (127) is drawn over the frames, and, without exposing the brood nest to cold winds, the dummy and the frames between it and the centre of the brood nest, are gently drawn back together a couple of inches; the dummy is then moved back a little farther, and the outside frame is lifted out (182). If it contain capped honey, the cappings are broken, and the frame is placed in the centre of the brood nest; the frames and dummy are closed up, and the sheets, quilts, and roof are placed in position as before. The operation may be repeated from week to week; frames of drawn out comb being used, until the brood chamber is filled with brood. In the absence of drawn out combs, frames of foundation may be used; but it is not advisable in the spring, when every day is of importance, to put upon the bees the necessity for comb building, thus delaying the increase of brood which might otherwise be expedited. Careful bee-keepers make it a rule to have always on hand a supply of drawn out combs for this purpose (317). Later in the season, and especially in the summer, combs or frames of foundation should be given in the brood nest as opportunity offers. (217).

"During summer, whenever a fairly strong stock is opened for any purpose (such as putting on or taking off sections) a frame should be put in the centre. This is a golden rule."—T. B. O'Bryen.

194. Drone Brood.—Towards the middle of May, when, in normal seasons, swarming may be expected, the bees construct drone cells (67). These will be readily recognised by their size, being deeper than the worker cells, and $\frac{1}{4}''$ in diameter, whereas worker cells are only $\frac{1}{5}''$ in diameter; the cappings of the former standing out beyond the cappings of worker cells (Fig. 14, F, page 36). The drone egg, like the worker egg, hatches in three days, and the grub is fed up to the ninth day from the laying of the egg; when, the cell is sealed, the spinning of the cocoon takes place, the change from larva to nymph, and on the twenty-fifth day the young drone makes his way out of his cell. About a fortnight later he leaves the hive for flight. (204).

BREEDING.

195. Controlling Drone Rearing.—The rearing of drones may be limited by the use of foundation prepared for worker brood, and may be encouraged by the use of drone-brood foundation **(113)**. It will, however, be found that, except in the case of a new swarm, if the former pattern be cut, or broken, or supplied in the frames as "starters" instead of in full sheets, drone cells will be attached to it in large numbers **(110)**. To avoid the extravagant rearing of drones, worker foundation is used in full sheets in all the frames of the brood nest, and the sheets are wired into the frames **(117)** so that they may not easily become sagged, or broken in the hive or extractor.

196. Queen Cells.—At the approach of the swarming season, if the queen and the rapidly increasing population of the hive become pressed for room; at any time when a colony has been deprived of its queen; or when the bees desire to supplant an old queen whose fertility has ceased, queen cells are started on the combs **(71)**. These are distinguished from all other cells by the material of which they are made, and by their size, shape, and position (Fig. 14, A, B, C, page 36). They are constructed of a mixture of wax and pollen; are about $1''$ long $\times \frac{1}{3}''$ in diameter, are in shape like an acorn, and they hang mouth downwards on the combs. The bees construct queen cells on the face of a comb by breaking down the cells immediately surrounding those containing the eggs from which queens are to be reared. At other times queen cells are made on the sides, or the bottoms of the combs; and, when the queen does not deposit eggs in them, bees have been known to carry eggs to them from other cells, lengthening the queen cells as the process of feeding the grub proceeds. The number of queen cells constructed by a colony of native, or black bees may vary from two to ten or twelve. Other races frequently exceed those figures. Syrian bees **(50)** will sometimes provide as many as thirty queen cells on one comb, and it is said that more than seventy queen cells have been found in one colony of Syrians. The cells are not all started on the same day, the object being to have the young queens hatch out in succession. In a case of emergency, arising when a colony has been deprived of its queen, if the bees have worker eggs available, or larvæ not more than three days old, *i.e.*, not already weaned **(190)**, they will construct a queen cell around the selected egg or larva. Should they have no worker egg, or larva under four days old, they will, in a desperate effort to retrieve disaster, form queen cells here and there at random, and even around drone larvæ. The latter cells, which may be distinguished from regular queen cells by their smooth walls (Fig. 14, G, page 36) cannot, of course, produce anything but drones—drones which, perhaps by reason of their too generous nursing, frequently die in their cells.

197. Nursing Queen Larvæ.—There is not any difference between the egg which produces a queen bee and that which produces a worker bee. But the treatment in the process of nursing varies considerably. The larva, in the former case, is given a cell which permits of its growth to the full dimensions of a queen, and is more liberally supplied with food—food of a richer quality, called "Royal Jelly" to distinguish it from the food provided for other larvæ: and, whereas the larva of the worker bee is weaned three days after it has left the egg, and is then supplied with a coarser food (190), the larva in a queen cell continues to receive abundantly the Royal jelly. Leuckart discovered that the development of the female genital organs begins upon the third day after hatching. This development continues under the liberal treatment referred to, and the produce is a mature female or queen; or ceases with the withdrawal of the stimulating food, when the result is an immature female, or worker. It follows that, for the production of a vigorous queen, the special treatment should begin with the egg, or at least before the larva has passed its third day.

198. Wonderful Effects of Special Nursing.—The effects upon the larva of this continued supply of richer food, are among the most wonderful in the history of bee life. The larva which, in the ordinary course of nature, we should expect to arrive at maturity by slower stages, reaches its full growth in about two-thirds of the time occupied by the worker larva (204). The young queen has her organs fully developed, so that, when fertilised, she can, during the ordinary span of queen life, produce impregnated eggs to the extent of 100 times her own weight; while the worker can never, by any means, produce an impregnated egg (200). In colour, shape, and size, she differs materially from the worker, being darker, more delicately formed, and with greater length (4). Her sting is longer, and curved (41). Her hind legs are without corbiculæ (34). Her abdomen is without wax secreting receptacles (37). Her eyes have only about 10,000 facets, as against the 12,000 facets of the worker (30). Her habits and instincts are, in many respects, the opposite of those of the worker—she confines herself to the duty of egg laying, never leaving the darkness of the hive after her wedding flight, except when accompanying a swarm: she is not disposed to sting even if molested by the bee-keeper: far from sharing the worker's deep-rooted reverence for the person of a queen, she shows a bitter hostility to all others of her own rank, and will fight to the death against a rival queen. She may live for four or five years; whereas the worker's life is limited to about six weeks, except in the case of workers born at the close of autumn, and surviving through the winter rest to labour for a few weeks in the spring. So marvellous are the developments brought about by the simple process of feeding.

199. Queen Brood.—The egg from which a queen is to be reared, like the egg which is to produce a worker, hatches in three days; for six days more it continues in its larval state; it then spins its cocoon, is transformed into a nymph, and, on the sixteenth day from the laying of the egg, it emerges a perfect virgin queen. The vacant cell is never employed again for queen rearing, but is cut down usually within a few hours (71) as shown, Fig. 14, C, H, page 36. Soon the young queen begins her search over the combs for a rival, and if permitted, she will destroy the unhatched virgin queens in their cells. (20). A few days later, if the weather be favourable, she leaves the hive for impregnation. (204-213).

"Hardly had ten minutes elapsed after the young queen emerged from her cell, when she began to look for sealed queen-cells. She rushed furiously upon the first that she met, and, by dint of hard work, made a small opening in the end. We saw her drawing, with her mandibles, the silk of the cocoon, which covered the inside. But, probably, she did not succeed according to her wishes, for she left the lower end of the cell, and went to work on the upper end, where she finally made a wider opening (Fig. 14, B). As soon as this was sufficiently large, she turned about, to push her abdomen into it. She made several motions, in different directions, till she succeeded in striking her rival with the deadly sting. Then she left the cell; and the bees, which had remained, so far, perfectly passive, began to enlarge the gap which she had made, and drew out the corpse of a queen just out of her nymphal shell. During this time, the victorious young queen rushed to another queen-cell, and again made a large opening, but she did not introduce her abdomen into it; this second cell containing only a royal-pupa not yet formed. There is some probability that, at this stage of development, the nymphs of queens inspire less anger to their rivals; but they do not escape their doom; for whenever a queen cell has been prematurely opened, the bees throw out its occupant, whether worm, nymph, or queen. Therefore, as soon as the victorious queen had left this second cell, the workers enlarged the opening and drew out the nymph that it contained. The young queen rushed to a third cell; but she was unable to open it. She worked languidly and seemed tired of her first efforts."
—*Huber.*

200. Laying Workers.—Although, as already stated (198), the worker bee is incapable of being impregnated, there are occasionally found, in a queenless hive, one or more workers whose ovaries, partially developed, contain a certain quantity of eggs (Fig. 11, C). Huber supposed that these laying workers were the produce of eggs deposited in cells adjacent to queen cells, and that they had received a smaller quantity of royal jelly. Possibly they are either workers which, in the early stage of development, were not weaned until after the third day, and whose organs have been partially developed by reason of the excess allowance of the richer food which they have received; or, workers which, in the larval state, were, at an age over three days, selected for special treatment by a queenless colony, and thus, in the earlier stages of the larval growth, were deprived of the liberal treatment necessary for the production of perfect queens (197). It is very rarely that a laying worker is tolerated in a colony which has a prolific queen. But, in colonies which are queenless, and which have neither eggs nor young larvæ from which to raise queens, laying workers are occasionally found, and sometimes in large numbers. Their eggs, being unimpregnated, produce drones only. Their presence in a hive is indicated by the irregular manner in which their eggs are deposited, several being frequently found in one cell, and cells with eggs appearing side by side with cells containing drone larvæ, whereas a fertile queen lays her eggs very regularly, as shown above. (Fig. 77, page 103).

201. Removing Laying Workers.—Laying workers must be got rid of, or the colony must perish. If a comb containing eggs from another hive be given to the colony, and if the bees can be induced to raise a queen, or queens, from those eggs, the laying workers will be destroyed so soon as a young queen begins to lay in the hive. But, where a laying worker has been in possession for some time, the bees of the colony are often indisposed to rear a queen from eggs supplied to them, and will refuse to do so while the laying worker remains in the hive. This difficulty may sometimes be overcome by altering the position of the hive for a few days; then removing all the bees, carrying them to a distance of 100 or 200 yards, and shaking them down there upon a sheet or board; when, the laying workers, unfamiliar with the new position of the hive, will fail to find it, while the other bees, except the useless young drones, will return to the hive, and will raise a queen from eggs supplied to them. Beside the fact that a colony long queenless will be short of, and perhaps destitute of nurse bees, this remedy entails a loss of some weeks before the young queen can begin laying, and of over two months before her progeny can supply the place of the dwindling workers of the colony; and it can be adopted (so far as queen rearing is concerned) only when there are drones flying to fertilise the young queen. The speediest, and the best remedy is to introduce a young fertile queen **(295)**. But, if a fertile queen cannot be procured, the colony may be united to another, or may be broken up, and divided among other stocks having fertile queens.

202. Stimulating in Autumn.—The bee-keeper, knowing that the success of the colony in the ensuing year will depend largely upon its going into winter quarters with a large supply of young bees, begins to stimulate again by supplying warm syrup, a little each evening, from the termination of the honey flow until about the middle of September **(314)**. In this way queen and bees are induced to keep up the numbers of the colony, and the danger of a scarcity of food is lessened.

203. Breeding Ceases.—When, with advancing autumn, the flow of nectar diminishes, the drones are destroyed **(24)**; the daily deposit of eggs by the queen lessens **(25)**, occupies fewer combs, and smaller circles, as the cluster of bees draws towards the centre of the hive; and finally, in November, or earlier if the season prove very inclement, it ceases altogether. In very mild seasons, and in a hive well supplied with stores, the queen will sometimes continue to lay well into December.

204. Metamorphosis of Bees.—The following data are supported by common experience of the metamorphosis, etc., of bees. But it must be understood that the figures and the dates given are only approximate, and are variable according to the strength of the colony, the heat of the hive, and the condition of the weather:

TIME OCCUPIED FROM THE LAYING OF THE EGG.

	Queen	Worker.	Drone.	
Incubation of the egg	3	3	3	Days
Feeding of the larva	5	5	6	,,
Cell sealed on the	9th	9th	9th	Day
Spinning cocoon	1	2	3	Days
Interval of inaction	2	3	3	,,
Change from larva to nymph	1	1	1	,,
,, from nymph to exit as perfect insect	3	7	9	,,
Bee evacuates the cell on the	16th	22nd	25th	Day
Interval spent chiefly in the hive	5	14	14	Days
Bee flies freely from the hive on the	21st	36th	39th	Day

Interval between issue of top Swarm and issue of 1st Cast ... 9 Days.
,, ,, 1st Cast and 2nd Cast ... ... 2 ,,
,, ,, 2nd, 3rd and 4th Casts ... ... 1 Day.

CHAPTER XX.

SWARMING.

205. Natural Swarming.—Natural swarming may occur upon any fine day from the middle of spring to the middle of autumn, as the condition of the colony may demand. As a general rule, swarms may be looked for from the end of May, or in a very favourable season, from the closing days of April, up to the termination of the honey flow. For profit during the season, early swarms are, of course, most in demand (152), because, if they come off in April or early in May, they have time to build combs and to rear brood before the opening of the honey flow; whereas if they issue in June or July, the honey flow, except in heather districts, will have ended before a sufficient number of young bees can be produced to take advantage of it (204); in which case neither the swarm nor the parent stock can store much surplus honey that season, and the former will probably require feeding and careful attention, to enable it to survive the autumn and winter. Hence the well-known adage:—

> A swarm of bees in May
> Is worth a load of hay;
> A swarm of bees in June
> Is worth a silver spoon;
> A swarm of bees in July
> Is worth a butterfly.

206. Signs of Swarming.—With the use of modern hives, the bee-keeper is able to calculate, with some degree of accuracy, the date upon which any particular stock is likely to swarm; and, by a little observation, he can avoid being taken completely by surprise. This is one distinct advantage of modern bee-keeping over the old methods; for, if any circumstance of bee-life demands more prompt attention than another it is the issue of a swarm, which must be dealt with at once, and which may be said, like "time and tide," to wait for no man. When a colony has increased in numbers to such an extent as to become cramped for room; when nectar is being carried in rapidly; and when drones are on the wing, preparations are made for swarming, and, seven or eight days before the event is to take place, queen cells are started upon the combs. The first of these cells will be sealed over on or about the ninth day, and, when this is observed; when the bees of the colony are found clustered about the entrance of the hive, or working in a listless, half-hearted way, while the bees of other stocks are actively engaged foraging—the swarm may be expected to issue. (See illus. p. 129.)

207. Delay of Swarming.—Should rain and unfavourable weather generally prevail at the time of the capping of the earlier queen cells, the swarm will not issue. The mature cells will be opened, the young queen nymphs will be destroyed, and swarming will be deferred until the weather improves; and, if necessary, fresh queen cells will be prepared, loss of valuable time ensuing. Should inclement weather be prolonged, swarming may be abandoned altogether for the season.

208. The Swarm.—But, if the weather continue favourable, the bees will, in the early forenoon, make ready for their departure. A number will be seen flying in front of the entrance, gaily sporting themselves, and with their heads towards the hive. Within, the agitated queen, having ceased ovipositing, hurries from comb to comb, where those of her progeny who are to accompany her in this reckless abandoning of home, and stores, and brood, are filling their honey sacs from the cells, laying in a supply of food sufficient to serve them for three or four days **(18)**. Presently wild excitement spreads through the whole colony; the bees rush hither and thither; the temperature rises rapidly; and, suddenly, the swarming bees pour out from the entrance in a steady stream. The air seems to be full of them; they fly around in the very abandonment of ecstacy; until, the queen mother joining them, or alighting upon some neighbouring tree, they settle around her, and form the well-known cluster of the swarm. Previous to this scouts have been sent out to find a suitable place in which the swarm may locate itself, and lay the foundation of a new home. Usually, until the return of the scouts (which may occur within an hour), the bees will remain in the cluster; and they should be secured at once, because the scouts generally select the new location at a considerable distance **(211)**, and, when the swarm rises from the cluster, it will follow the scouts, and may be lost to the owner. **(19).**

209. Vagaries of Swarms.—Should the queen, from any cause, fail to leave the hive, the bees will return, and will endeavour to force her to accompany them; for, they will not venture upon this hazardous enterprise without their mother bee (16). Should the queen leave the hive, and fall to the ground, the bees which discover her will cluster there, and the remainder will return to the hive. Should the swarm, when clustered, disclose a disproportion of young or of old bees, the swarm may go back, and may issue again; and this may be repeated several times, until the proper proportions are arrived at. Should the scouts fail to find a suitable location before sunset; or should rain suddenly appear, the swarm may remain in the original cluster until the following day. If, on the other hand, the weather be hot, and if the cluster be left unsheltered from the sun, the swarm may decamp at once, without awaiting the return of the scouts. Sometimes the queen, a stranger to light, and unaccustomed to fly, is unable to reach the selected spot, and will drop, exhausted, on the way, and the new home be started in an unsuitable place. Occasionally two swarms come out at the same time and form one cluster; these should be treated as one swarm, and, on being hived, one of the queens may be removed for use elsewhere. If both queens be allowed to enter the new hive, one of them will be destroyed. (See also **185b** and **254c**.)

210. To Encourage Clustering.—The old-time custom of beating tin cans, in order to cause the swarm to settle quickly, is possibly due to its having been observed that flying bees hasten home from the fields when thunder storms threaten in summer; and, the din one sometimes hears in swarming time is intended to represent the "artillery of the gods." Bees are highly sensitive to the approach of rain, and will seek the shelter of their hives when rain is near. But, it is very probable that it is not the thunder which may precede a summer shower that influences them; and it is not likely that the noise of horns and drummed cans can have much, if any, effect in causing them to cluster rapidly. Water, however, may be used with good effect. If applied through a garden syringe which casts a fine spray, and so that it fall upon the swarming bees from above, like rain, it will hasten their settling, and will cause them to cluster closely, so that they may be the more easily, and the more promptly secured. It is recorded that truant swarms have been headed off, and impelled in the required direction by this means.

211. Truant Swarms.—Swarms, when they once rise from the first cluster, seldom remain in the vicinity of their former homes (19). It appears to be their object to settle as far as possible from the hives which they have abandoned, and to leave to their successors not merely the stores there, but also the flowers of the immediate vicinity. This is one of nature's provisions against the mischief of in-breeding. It is often a cause of disappointment and loss to the owner, who tries, in vain, to stay or to overtake his truant swarm. The law, as it applies to the ownership of truant swarms, seems to be, that if the bees have been seen issuing from their hive, and have been kept in sight by the owner, or by someone on his behalf, while they have been followed, and until they have entered the premises where they cluster, they may be legally claimed and removed. Otherwise, they become, in the eyes of the law, *feræ naturæ* or wild bees, and may be claimed by anyone who takes possession of them.

212. Clipping Queen's Wings.—Many devices have been employed to induce swarming bees to cluster in accessible places, and to counteract the instinct which impels them to depart to "fresh woods and pastures new." Decoy hives, furnished with some combs, will sometimes entice swarms to take possession of them. Dry, dark combs, and even black hats and stockings, tied to the lower branches of trees in the apiary, are said, by reason of their resemblance at a distance to clustering bees, to have an attraction for swarms. But in spite of every such device, swarms will frequently cluster in the high branches of trees, or in other inaccessible places, and they may decamp altogether before they can be secured by the owner. The difficulty may be prevented by the simple expedient of clipping the queen's wings; for, if the queen cannot fly, the swarm will not decamp; and if it should settle upon a high branch, it will, when the absence of the queen is discovered, return to the hive. Accordingly, if the queen be picked up, she can be allowed to run in with the bees when the swarm returns to the hive (either the parent hive or a new hive placed on the old stand), and thus the trouble of following and securing the swarm may be obviated. The proper time for clipping is in the early spring when the population of the hive is small, and when, therefore, the queen can be more readily found. To clip a queen's wings, proceed as follows:—Take out the frame on which the queen is found, drawing the carbolic cloth over the brood nest, and rest a corner of the frame on the hive: follow the queen with a small scissors as she moves about, and watch your opportunity to pass a blade of the scissors under the larger wing on one side, and clip off a portion of it (Fig. 82),

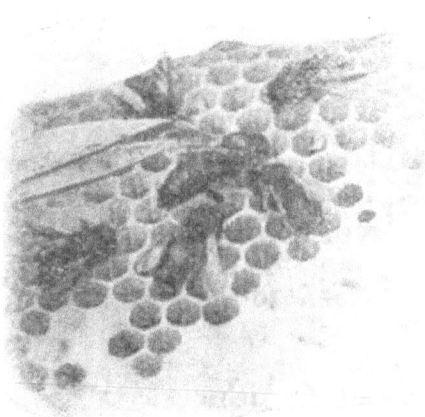

Photo by]   [*J. G. Dinges.*
Fig. 82.
CLIPPING QUEEN'S WING.

Another method may be followed:—Hang the frame, upon which is the queen, upon a comb stand (172) and pick off the queen by the wings with the finger and thumb of the right hand, as shown (Fig. 83, A); then gently take her, by the thorax, in the fingers of the left hand, clip the wing (Fig. 83, B), return her to the brood nest, and place the frame in its former position in the hive. In either case the operation

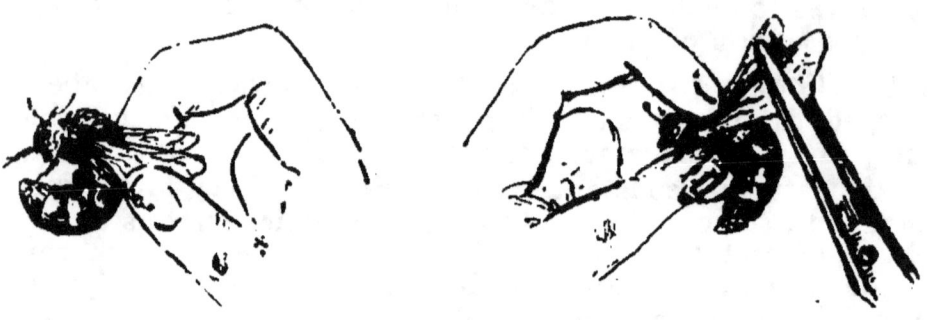

A   Fig. 83.   B
CLIPPING QUEEN'S WING.

is a delicate one, and should be carefully performed, avoiding all risk of injuring the queen by any pressure upon her abdomen. By this means, also, the ages of queens may be recorded upon their persons; the wings on one side being clipped in their first year; those on the other side, in their second year; and, in their third season, when there are no longer any wings to clip, a young queen should be given to the stock (281). It goes without saying that queens should not have their wings clipped before they have been mated. The risk of losing swarms is avoided also by Artificial Swarming (222).

213. The Parent Stock.—Seven or eight days after the issue of the prime swarm, the first of the virgin queens emerges from her cell, and, if the stock decides against further swarming, the young queen, assisted by the bees, destroys the royal nymphs, and assumes her position as queen of the colony (199). About five or seven days later, *i.e.*, thirteen to sixteen days after the issue of the prime swarm, she leaves the hive for impregnation, and, usually on the twenty-first day after the swarm, her eggs may be found in the cells. Although those dates are only approximate, they are reliable enough to guide the bee-keeper in his management. He will know, for example, that, on the twenty-first day after the swarm, all worker brood of the old queen will have emerged from the cells, and that the young queen will have only just begun to lay. If, therefore, he desires to transfer bees and combs from skeps to modern hives (253) he will select the twentieth or the twenty-first day after the skep has given a swarm, as offering least risk of injury to brood.

214. Casts.—As above stated **(206)**, swarms usually issue shortly after the sealing of the earliest queen cell, which takes place on the ninth day from the laying of the egg. Seven or eight days later the young queen leaves her cell, and attempts to destroy the royal nymphs **(199)**. Early in the morning, or in the afternoon, when the bees are still, her shrill piping may be distinctly heard, and also the muffled, piping answers of the royal nymphs who, still imprisoned in their cells, are conscious of impending danger. If the colony be sufficiently strong to give off a cast, the bees mount guard around the queen cells, and refuse to permit the young queen to destroy her rivals. On the following day, which usually is the ninth day after the departure of the prime swarm, the young queen and the second swarm, or cast, issue **(20)**. If, however, the weather be very unfavourable, the exit of the cast may be delayed, and even the queen cells and their occupants be destroyed (Fig. 14, B, page 36), and further swarming be deferred, or ended for the season. But young, unmated queens are somewhat reckless and impetuous, and will often come out with a cast on a rainy day; they will fly farther than aged queens before alighting; and such casts are more likely than are prime swarms to abscond, even after they have been hived. Second casts usually issue two or three days after the first cast, and third and fourth casts on the next and the following days respectively. Sometimes two or more virgin queens, emerging from their cells on the same day, accompany one cast. The cast being hived, all but one queen will be destroyed. Although, in a favourable season, an early first cast may be made profitable, after swarms should be discouraged because, weak themselves, they so depopulate the parent stock that neither can be of much use that season.

215. Hunger Swarms.—Occasionally bees will forsake their hives on any day of the year, except in winter, either as a complete stock or as a swarm, and will locate themselves in any available nook or corner. This may arise from the presence in the hive of something distasteful to the bees **(232)**. But most frequently it is the result of hunger; when, a portion of the bees will abandon the hive, in a spirit of self-denial leaving such food as remains for the queen and the remainder of the stock; or the whole stock will depart, knowing that starvation is imminent, and in a desperate hope of bettering their condition elsewhere **(307)**. Obviously, the remedy is, in the former case, to introduce them to a clean hive; in the latter case, to provide them with food. Examination of the hive will generally show what has been the cause of the departure of the bees, and, when the cause has been removed, the truants may be returned to their old quarters.

216. Prevention of Swarming.—It is frequently advisable to prevent natural swarming, because of the trouble and risks attending it, and because, when one desires to obtain the largest possible harvest of honey, and does not wish to increase his stocks, natural swarming upsets all his arrangements; for, it is not possible, in an average season, to secure both an increase of stocks and a large supply of surplus honey. Even where increase of stocks is chiefly desired, natural swarming may be prevented with advantage, and, by artificial swarming the increase be made by wise selection from the best colonies **(222)**. It must be remembered that it is, generally, quite impossible to prevent swarming when once the bees of a colony have contracted the "swarming fever" **(187)**—so-called, perhaps, because like any fever that "flesh is heir to," when once it has set in, the arrival of the crisis is inevitable. Therefore, the bee-keeper, desiring to prevent natural swarming, and familiar with the causes which promote it **(206)**, should set himself, in good time, to circumvent them.

217. Giving Room.—One fruitful cause of natural swarming is congestion in the brood chamber, when there is not sufficient room either for ovipositing by the queen, or for honey storing by the bees **(187)**. Therefore, before they are actually needed, frames of comb, or of foundation should be added to the brood nest, and, in the season, new sections or frames to the supers **(255)**. When the honey flow is on, *i.e.*, when nectar is being carried in rapidly, the addition of frames of foundation will not always meet the needs of the case; because, the demand for vacant cells, both for eggs and honey, becomes too urgent, and, before the foundation can be drawn out into cells, congestion may set in, and preparations for swarming begin. In such circumstances, empty combs should be given. If the hive has already its full complement of frames, one or two frames of honey may be removed, the honey extracted **(134)** and the frames returned to the hive; and this should be repeated weekly, or more frequently, as required. In a pressing case, one or two frames of brood may be removed and given to another stock, the vacancy being filled with empty combs. Thus, not only is the tendency to swarm checked, but the storing of honey is largely increased. Afterwards, when supers are put on, the pressure upon the combs in the brood nest is relieved, and if, as each fresh super is added, one or two of the frames in the lower storey have their honey extracted, and are returned to the centre of the brood nest, or if a frame of foundation be given there, the queen will have sufficient scope for her energies below, the stock for their energies above, and the inducement to swarm will be minimised, if not entirely removed. (**193** and illus. page 129.)

218. Ventilation.—Excessive heat in a crowded hive encourages swarming. Therefore, hives, in warm weather, should be well ventilated; the doors should be opened to full width; the ventilator in the floor board (**85**) should also be opened. A ventilating dummy (**95**) may be used at the back of the brood chamber, the body box being moved backwards on the floor board to admit air through the dummy, or an opening in the back of the hive being provided for that purpose, so that it can be closed from the outside, or partially closed as required. The floor board, when constructed so as to admit of this, may be lowered, to admit air from all sides. The body box may be raised half an inch from the floor board by wedges at the corners. The roof may be tilted up in the front, may be shaded from direct sunrays by trees, or by a make-shift shade of one kind or another; in extreme cases, a sack may be soaked in water and placed upon the roof, and be kept damp and cool during the hottest hours of the day. Bees will not for long tolerate an upward draught, and, although to meet a sudden emergency an upward draught may be caused by placing a feeding stage (**121**) upon the sheet and raising the roof, such an expedient must be only temporarily adopted. Hives should never have the floor board permanently fastened to the body box, because of the difficulty, among others, of ventilating; and all ventilators applied to a hive should be of such a nature as to be easily opened, and as easily closed, without the risk of disturbing or crushing bees; for, in our climate, chilly nights frequently follow warm days, when, should the ventilators be left fully open, chilled brood might result. (**338**).

219. Limiting Drone Rearing.—A third circumstance incident to swarming lies in the breeding of excessive quantities of drones—fussy, and somewhat pushful insects which raise the temperature of the hive, and by their very presence suggest, continuously, the rearing of young queens. Therefore, the production of drones should be limited by the use of only worker-cell foundation, and in full sheets, wired, to prevent breakage and consequent construction of drone cells (**195**) and also by cutting out unnecessary drone comb when discovered in the hive. The skilful bee-keeper makes it a point to limit drone rearing in all his stocks, except in those that are headed by his best queens. Thus he secures that the drones which shall fertilise his young queens shall be of the best blood in his apiary, and by careful selection he keeps up, and even improves the quality of his stocks.

220. Limiting Queen Rearing.—A fourth condition in the stock which is about to swarm, is the presence of queen cells; and it is sometimes recommended, as a preventive of swarming, to cut out all such cells. The advice is based upon the theory that it is the mother-queen who causes the swarm by stirring the bees up to it, and by leading them out; that she is impelled by her wrath at the rearing of young queens; and that, if the queen cells be destroyed by the bee-keeper, the old queen will be placated, and the swarm be prevented. But it is evident to careful observers that swarming generally takes place, not at the instigation of the old queen, but against her will; that she does not lead the swarm out; and that she frequently shows a pronounced disinclination to leave her hive, and, sometimes, has to be driven out by the bees. The swarm becomes necessary by reason of the conditions referred to above; the bees prepare for it in the manner already described; and, cutting out the queen cells, while it will delay the swarm, will not alone prevent it: further queen cells will be formed; the preparations will be continued; and, the bees, always listless during such periods, will sacrifice much valuable time and energy in those weeks of the year which, to the bee-keeper, are most valuable. Therefore, while cutting out the queen cells may be adopted in connection with the other preventive measures, it will not, by itself, accomplish the desired object. Sufficient room and ventilation must be given, and it should be remembered that, if the various precautions be deferred until the bees have felt the need of more room, the swarm will probably issue in spite of all that the bee-keeper may do. **(216).**

221. Prevention of Casts.—Casts may generally be prevented by removing all queen cells from the swarmed stock, giving extra room, and introducing a fertile queen. **(240—242).** If a cast be hived on one or two frames of brood, it will quickly increase to a good stock.

222. Artificial Swarming.—As a substitute for natural swarming, artificial swarming, in the hands of a capable bee-keeper, offers many distinct advantages. It enables him (1) To arrange, by careful selection, the increase of his stocks, and that, always from his best queens: (2) To obtain early swarms, and from stocks which, if left alone, might not swarm naturally: (3) To prevent serious waste of time by stocks in preparation for natural swarming: (4) To avoid excitement and trouble securing and hiving swarms, and the risk of their absconding altogether **(211)**: (5) To provide swarms for sale, as required **(227)**: (6) To introduce strange queens to stocks **(298)**: and (7) To remove bees from infected combs, as in the treatment of foul brood. **(356).**

223. Conditions.—There are certain conditions which require attention in all the following operations, viz.:—(1) The stocks to be operated upon must be strong. (2) There must be drones hatching, or on the wing, to fertilise the young queens. (3) The day must be fine, so that there may be sufficient flying bees to form the swarms. (4) The brood must be carefully protected from cold. (5) The swarms must be fed for a few days, especially if they have no sealed honey in the combs given them.

224. One Swarm from One Colony.—From the stock to be swarmed remove a frame of brood with the queen and adhering bees, and place it in a new hive. Add, say, six frames of comb, or of foundation, three on either side of the occupied frame: close up the dummy: put on the quilts and roof: and set the new hive upon the stand of the parent stock, removing the latter to another position at least six feet away. All the flying bees of the parent stock, returning to their old stand, will form the swarm. The parent hive should be given a frame of comb in the place of the frame removed; not a frame of foundation, if comb can be procured, because queenless bees are disposed to build cells suited to the storing of honey rather than to the rearing of workers, and this is to be avoided. If a fertile queen be introduced **(295)** to the parent stock in the evening, no time will be lost in brood rearing. If a fertile queen cannot be supplied, one or two ripe queen cells may be given. If neither queen nor queen cells be available, the bees will rear a queen for themselves.

225. One Stronger Swarm from Two Colonies.—Prepare a hive (S) with seven or eight frames of wired foundation. Remove a strong stock (A) to one side, and place the hive S on A's stand. Set a hiving board, with a white cloth upon it **(233)** and sloping from the ground to the alighting board of hive S. Smoke the bees of hive A, and take out the frames one by one, brushing, shaking, or thumping **(184)** the bees on to the hiving board, until all have been removed from hive A, when they will run into hive S, and will be strengthened by the addition of all the flying bees of A returning to their old stand. The frames of hive A having been returned to their hive, and kept covered, to avoid the danger of chilling the brood **(338)**, remove a second strong stock (B) to a new position, and place hive A upon B's stand. All the flying bees of hive B, numbering many thousands, will enter hive A, to rear the brood and to raise a new queen for themselves. If, in the evening, a fertile queen be introduced to A **(299)**, breeding will proceed without interruption, and much valuable time will be saved. By this method an extra strong swarm is secured without unduly reducing the strength of the two stocks operated upon.

226. Using Three or More Stocks.—When there are more than two stocks available for the purpose, the above method may be varied as follows:—Remove one strong stock to a new position, and place an empty hive upon its stand. Take, as required, one, two, or more frames of brood from the other stocks, returning the adhering bees to their hives, and insert the frames in the new hive, supplying their places with frames of comb, or of wired foundation. Thus the first stock supplies the bees, the others the brood, and none of them is appreciably

227. Making Swarms for Sale.—When swarms are being prepared for sale, they may be made up from one, or more stocks, as desired. If from one stock, the frame on which the queen is found is removed, and the bees upon it, with the queen, are brushed, or shaken into a swarm-box **(160)**, or upturned skep, and as many more bees as are required are also shaken in. If still more bees be required, the box, or skep, may be placed upon the stand of the parent hive until a sufficient number of flying bees have entered it. It may then be prepared for transit **(153)**. Another method is to set the empty skep temporarily upon the stand of the parent hive; a frame is then removed from the parent hive with the queen and some bees; the queen is picked off the frame and placed at the entrance of the skep, and the bees are shaken off the frame so that they may run in with the queen: the operation is continued with other frames until sufficient bees have been transferred to the skep, which is then prepared for transit, the parent hive being returned to its stand. If bees of more than one colony are required, care must be taken to include only one queen in the swarm; and the bees should all be dusted with flour, or aspersed with thin, scented syrup to prevent fighting. **(160)**.

228. One Swarm from a Stock and a Nucleus.—It will be evident that one of the objections to swarming, both natural and artificial, lies in the fact that the stock which has been deprived of its queen, while the older bees are dying off rapidly, must be without a laying queen for at least twenty-one days **(213)**. Careful bee-keepers overcome this objection by having a supply of young, fertile queens in nucleus hives **(290)**. Where such queens can be had, artificial swarming may be carried out without any waste of time, and therefore more successfully. The following procedure may be adopted:— Upon a fine day, when honey is coming in, secure the young laying queen of the nucleus upon one of the frames, by a pipe-cover cage **(297)**, so that she shall have some honey at her disposal. If the nucleus is in a small hive, transfer the bees and combs to a suitable hive, and add sufficient frames of comb or of foundation. Place the hive containing the nucleus on the stand of a strong stock, removing the latter to the stand formerly occupied by the nucleus. By this method the nucleus receives the flying bees of the stock; both colonies have fertile queens; the risks attached to other methods are avoided; and the bees will work with a will. The caged queen may be released in thirty-six hours.

229. Making Swarms from Stocks in Skeps.—Owing to the inconvenience of feeding and observing bees in skeps, and to the difficulty of supplying swarms in skeps with brood and drawn-out comb, artificial swarming to colonise skeps should be deferred until the stocks are not only strong, but also near the swarming stage, and until the weather is good, and there is abundance of nectar to be gathered. Drive (160) the stock from which the swarm is to be taken until the queen and about half the bees have gone up. If more than the desired quantity of bees pass up before the queen, pick up the queen and place her in a small box for the moment. Throw the excess bees back among the combs, and liberate the queen among the bees in the upper skep. Put back the parent stock upon its old stand, and the driven swarm upon a new stand, or *vice versa* as your object may demand. If the queen has not been found when all the bees have been driven, the stock must be returned to its skep, and the operation be repeated later on.

230. A Stronger Swarm from Two Stocks in Skeps.—When a stronger swarm is desired, and when two stocks (A and B) in skeps are available for the purpose, drive the queen and all the bees from stock A: place the driven swarm (S) upon A's stand: place stock A upon the stand of stock B: and remove stock B to a new position. Thus A gives all its bees to S, retaining the combs and brood, and securing, on its new stand, all the flying bees of B, while B is not depleted beyond its force of flying bees.

Photo from life] [by J. G. Digges.
CONGESTION (217). BEES CROWDED OUT.

CHAPTER XXI.

HIVING; UNITING; AND TRANSFERRING BEES.

231. Confidence in Protection from Stings.—It has already been stated that bees, when swarming, are most peaceable, and that at other times they may be "subdued to settled quiet" so as to be comparatively harmless **(167)**. But, for the reasons mentioned **(169)**, the beginner will do well to wear a veil and gloves when preparing to hive a swarm; for, until he gains the confidence which follows experience, the confidence arising from the feeling of being, for the occasion, protected from stings, will assist him to carry through the work in hands in a business-like way.

232. Preparing the Hive.—Some days before a swarm is expected, a hive should be prepared to receive it. If the hive has been used before, it should have any necessary repairs, and two coats of good paint. The inside, and the dummies, should be scalded, and washed with a solution of 1 oz. Calvert's No. 5 Carbolic Acid to 2 oz. water, or 1 teaspoonful Izal to 1 quart water, and the parts should be set out in the air so that the smell of the carbolic may disappear before the hive may be required, for, any unpleasantness in their new home might cause the bees to forsake it **(215)**. The hive, with eight or nine frames of comb, or of wired foundation, should then be set up in the position which it is to occupy in the apiary, carefully levelled, as previously directed **(147)**, and with the sheet and quilts upon the frames. The sheet, if new, should be soaked in water, and put, while still damp, upon the frames; it will then lie perfectly flat, and will continue to do so when dry. A solution of 1 oz. Yadil to 3 oz. water may be used instead of either of the above-mentioned disinfectants.

233. Hiving Swarms Direct.—When a swarm issues, no time should be lost in securing it. If a garden syringe is at hand, spray some water over the bees, and when they cluster, give them some more water to cool them, and to cause them to cluster more closely. If they cluster upon a low branch, or shrub, bring the prepared hive as close as possible to the cluster: place a hiving board sloping up to the alighting board of the hive, and raised at the other end so that it may be nearly, but not quite level: cover the hiving board with a white cloth arranged to lie smoothly right up to the hive entrance, and kept in position by stones at the corners: draw

HIVING BEES.

out the hive doors, and with them wedge up the front of the hive an inch or two from the floor board. If the swarm hangs not more than a foot or two above the hiving board, give the branch on which it hangs a smart shake, throwing the bees on to the white cloth. If the branch be too high, cut it off, without disturbing the cluster, and shake the bees on to the white cloth. They will speedily run into the hive; when, if a sharp look out be kept for the queen, she may be seen passing in. Should the bees delay to enter the hive, take up a handful, and place them at the entrance; or, with a feather move them on. The "music" which they will make on discovering their new home will act as a "quick march" to the remainder, and the swarm will soon be hived. When the bees are in, throw the white cloth over the hive, letting it hang down in front, thus sheltering from the sun until the flying bees join the others, when the hive must be carried to its permanent stand, for, otherwise, the bees, beginning work, will mark the spot, and will return there after the hive has been moved. Feed for a few days. (236).

234. Swarms in High Trees.—If the swarm should settle on a very high branch, the hive need not be removed from its stand. Get a clean skep (78), mouth upwards, under the cluster, and shake the swarm into it. If you cannot reach high enough, hold up the skep on a pitch fork, and get someone to shake the bees in; but take care to have your skep so secured that it will not topple over, and drop the swarm upon your head. The job may be more satisfactorily carried through if you provide yourself beforehand with a half sack, arranged on a hoop, and with a long handle, after the fashion of an angler's landing net. Get the sack under the swarm; shake the bees in; give the

SWARM IN A HIGH TREE.

handle a turn, closing the mouth of the bag so that no bees can escape, and carry them to the hive. If the branch must be cut off and removed, with any risk of shaking off the swarm in the process, bore two holes at opposite sides of the skep mouth and pass two strong cords through the holes: get the skep under the cluster and tie it to the branch: pass a sack up, covering the skep and swarm, and tie it also to the branch: then, no matter what jarring or shaking may ensue, the bees will be secure. Doctor Smyth's Swarm Catcher (fig. 84) is a device for securing swarms when they locate themselves in the branches of tall trees, or in other awkward places. It consists of an arched piece of ½" iron rod on a long pole. The ends of the arch are connected by a bar passing through holes at its extremities, and locked, when necessary, by a thumb screw. The bar carries two laths, fixed so as to form carriers for four or six frames. A thin lath thrust under the bar and pressing on the tops of the frames, holds them so tightly that, while the frames swing freely on the bar, they are fixed to one another. Thus an attractive temporary home may be brought within reach of a clustering swarm, and in it, without the inconvenience of climbing and branch cutting, the bees may be secured and conveyed to a permanent residence.

235. Swarms in Awkward Places.— Should it be impossible to work any of the above plans: if the swarm has entered

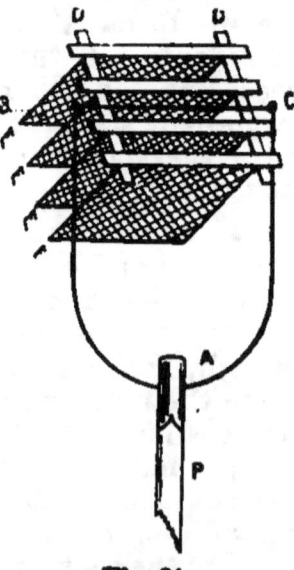

Fig. 84.
THE SMYTH SWARM CATCHER.

a chimney, or has clustered in a thick hedge, or in an old wall; get your skep over it, and drive the bees up with smoke, or with the fumes of a cloth saturated with carbolic solution (127). If you can so arrange that the smoke or the carbolic fumes will not enter the skep, the bees will march up. If the swarm clusters on a wall, or on the trunk of a tree, get the skep under, and with a brush sweep the bees in. If they alight on the ground, as they may possibly do when the queen's wings have been clipped (212), place the skep beside them, and raised an inch or two upon a couple of stones; with a twig, or a feather, or your hand move some of the bees to the entrance: they will pass in, and the remainder will quickly follow.

236. Hiving from a Skep.—Having secured the swarm in your skep, set the skep on a cloth or on the ground close to the place where the swarm originally clustered, and raised upon a couple of stones, to allow the outside bees to enter: cover with a cloth, to shield from hot sunshine, until all the bees have gone in: then carry it gently to the stand which it is to occupy. In the afternoon, a couple of hours before sunset, hive the swarm as directed (233). Or, if it be desirable to adopt another and more rapid method, remove the sheet and quilts from the hive which is to receive the swarm; space out the frames as widely as possible; and arrange the hiving board and white cloth as directed above (233). Take the skep; invert it; give it a good "bump" upon the ground to loosen the foot hold of the bees; and pour, or shake some of the swarm on to the frames, and the remainder on to the hiving board. Then spread the sheet, only, upon the frames; put on a feeder (119) with thin syrup (Recipe 321); and place the roof in position. In the morning, close up the frames and dummy, first removing any unnecessary frames, and any in which the foundation may have broken down; remove the wedges, and lower the hive front to the floor board; put on the covering and the roof; and continue gentle feeding for about a week.

237. Secure all the Cluster.—If, by any means, you have failed to secure the queen with the swarm, the bees will forsake the hiving skep and will return to the original cluster if the queen be still there, or to the hive from which they swarmed, in the event of their being unable to find the queen. Therefore, care should be taken to secure all the bees of the cluster, so that the queen may not be lost.

238. Sweetening the Hiving Skep.—The old-fashioned custom of smearing the skep with treacle, butter, or beer, as an inducement to the bees, is both useless and objectionable; but, a little piece of comb, with honey or brood, fastened by a skewer in the top of the skep, serves as an attraction.

239. Hiving by Caging.—If the bee-keeper is on the alert and sees the swarm as it issues, he may often save himself a great deal of trouble if he watches the queen as she comes out on the alighting board, and slips a pipe-cover cage **(297)** over her. He then places the prepared hive on the stand of the parent stock; sets the queen, in her cage, on the alighting board of the former; and waits for the swarm to return, when the absence of the queen has been discovered **(209)**. He then releases the queen, allowing her to go in with the swarm, and either leaves the swarm on the old stand, or removes it to a new position, as his requirements may suggest. If left upon the old stand, it will receive a large accession of strength from the flying bees of the parent stock, and will work with surprising vigour.

240. Hiving a Swarm on the Old Stand.—If you are working for honey rather than for an increase of your stocks, place the swarm upon the stand of the parent stock, removing the latter to a new position, and transfer the supers, if any, from the stock to the swarm. Thus, casts will be prevented; the swarm will be strengthened by the flying bees of the stock, and new energy will be thrown into its work. An excluder under the supers will be useful. If, at the close of the honey flow, you unite the two stocks, removing the old queen, in the following year you will have a strong stock, with a queen in her prime.

241. The Heddon Method.—The Heddon method is to move the parent hive to one side, beside, and at right angles with its former position, the hive with the swarm being placed on the old stand. Two days later, the parent hive is turned round so that its entrance points in the same direction as the entrance of the hive containing the swarm; and, seven or eight days after the issue of the swarm, *i.e.*, a day or two before a cast might be expected **(214)**, in the middle of the day, when bees are flying freely, the parent hive is changed to a new position, thus giving all its flying bees to the swarm, and effectually preventing casts. **(221)**.

242. Returning Swarms.—Another method consists in taking away all brood from the swarmed stock, filling the vacancies with frames of wired foundation, and transferring the brood at once to other hives; after which the swarm is run into the parent hive as directed **(233)**. Thus the swarming impulse is usually satisfied, the bees are kept together, and the foraging propensity receives a new stimulus. It is to be noted that the brood combs, before being given to other stocks, should have their queen cells removed.

243. Retracing Swarms.—Should there be any difficulty in locating the hive from which a swarm has issued, take from the cluster a handful of bees: put them into a small box, and dredge them with flour: then carry them to a distance, and shake them out upon a board or newspaper. If a watch be kept upon the hives, the bees that have been floured will be seen returning to the hive from which they issued with the swarm.

244. Uniting Bees: Precautions.—Weak stocks can never be profitable; but, if two or more of such stocks be joined together they will, in summer, do useful work, and in winter they will consume less stores, preserve their heat better, and will survive where, separately, they would perish. Frequently it is desirable to unite stock to stock, swarm to stock, or swarm to swarm, as the case may be. It must be remembered that bees of different colonies will not usually unite peaceably, unless precautions have been taken to prevent their fighting. Such precautions should aim at—(1) Causing the bees to fill themselves with sweets; and (2) Giving them the same scent.

245. Uniting Swarms.—Swarms, however, being already well filled with honey **(208)** and having neither home nor brood to defend, may be united at once if they be thrown together into one skep, or on to a hiving board, and allowed to run into the hive. One queen may be removed, or the two queens may be left to settle their differences in their own way. (See **254c.**)

246. Uniting Two Stocks.—Bring the two stocks (A and B) together as already directed **(156)**. Begin the operation of uniting in the evening, when all the bees have returned to their hives from the fields, because, bees entering after the union has taken place, and not having the same scent, may be attacked and killed. If a spare hive (C) be available, place it between the hives A and B. Subdue the bees by smoking them **(177)**, and, if they have no stores from which to fill themselves, give them some thin, warm syrup **(181)**. If one queen is better than the other, take away the latter. If hive C will not hold all the frames of the hives A and B, reject, for the present, the outside frames, and such others as have no brood, or are least valuable. From hive A take out frame No 1: thoroughly dust all the adhering bees with flour, and place the frame, with its bees, in hive C, and in a similar position to that which it occupied in hive A: take out frame No. 1 from hive B, flour the bees upon it, and place it in C, next to the former frame: proceed similarly with frames 2 A and 2 B, and continue until all the frames necessary have been transferred. Dust with flour all the bees remaining in A and B, and shake them, or brush them on to the frames in C. Cover up with the sheets, quilts and roof: give two or three puffs of smoke at the entrance, two or three thumps with your fists upon the roof, and "leave well alone" until the next day. If a spare hive be not available, space out the frames in A or B, and arrange the frames alternately in either hive.

247. Uniting Queenless Bees to a Stock.—In this case, protect the queen by caging, as directed **(297)**, and proceed as before **(246)**. If there is nothing to be gained by transferring all, or any of the frames from the queenless colony, the bees may be shaken into a skep, and thrown down upon a hiving board before the hive of the stock to which they are to be united. Another usually successful method is to cover the frames of the queenless stock with a sheet of strong paper, slightly damped, and having a few small holes pierced in the centre, then setting the hive with the other stock on top, and leaving all undisturbed for at least four days.

248. Uniting a Swarm to a Stock.—Proceed as described above for uniting queenless bees to a stock; with this addition, that, if you wish to preserve the queen not of the swarm, but of the stock, the queen of the swarm should be removed, because, otherwise, the two queens will fight, and if the queen of the swarm has not been fertilised, being the queen of an after-swarm, or cast, the fertile queen will probably be killed. If you cannot find the queen otherwise, allow only a few bees to enter the hive, keeping the bulk of the swarm well back from the entrance: then place a piece of excluder zinc **(109)** over the entrance: pick up the queen as she endeavours to pass through the zinc: remove the zinc, and let the swarm go in. Sometimes, to further reduce the risk of fighting, it is preferred to mix the bees more thoroughly by shaking those of the stock also on to the hiving board, allowing them to run in with the swarm.

249. Uniting Driven Bees.—Driven bees **(160)** may be united without difficulty. Dust the two lots thoroughly with flour. If they are in skeps, bring the skeps together, mouth to mouth: give them a "bump" on the ground, to throw the upper bees into the lower skep: shake, and mix them well together, and throw them on to the hiving board. If you do not remove one queen, the bees will settle that matter for themselves.

250. Uniting Driven Bees to a Stock.—For this operation, extra precautions against fighting are desirable. Procure a second, temporary, hive, and to it transfer about half the frames from the stock hive, returning the adhering bees to the latter as you proceed, and, without delay, hive the driven bees by shaking some on to the frames and the remainder on to a hiving board **(236)**. Now bring the two hives close together. After three or four days, unite the two lots as described under the head of "Uniting Two Stocks." **(246)**.

251. Transferring Bees.—When "spring cleaning" **(384)** is being attended to; when it is desired to change from the old to the modern methods; and at other times, it is found necessary to transfer bees from one hive to another, or from a skep to a modern hive.

252. Transferring from Hive to Hive.—Remove the stock hive to one side, and set a clean, empty hive on the vacant stand, with its floor board perfectly level, if the frames are to hang parallel with the entrance **(147)**, and with the frame carriers vaselined **(174)**. Subdue the bees with smoke: remove the quilts and sheet: if there are supers on, set them on two sticks on the ground, or upon a table: draw back the dummy, and space out the frames. Take out the frames, one by one, and insert them, in the same order, in the clean hive: set a hiving board in front, and brush or shake on to it any bees remaining in the old hive: replace the supers (if any) and put on the covering and roof **(384)**.

253. Transferring from Skep to Modern Hive.—The transfer of combs from a skep to the frames of a modern hive, is not often desirable. It is a messy, troublesome job that often leads to chilled brood **(338)**, and the combs are frequently worked out so irregularly in the frames as to render subsequent manipulations very difficult. It is generally preferable to allow the bees to transfer themselves by the automatic method to be described below **(254)**. But, where transfer of the combs is decided upon, the following process may be adopted:—Drive all the bees out **(160)**, and set them, in their skep, on their old stand. Take the old skep to a warm kitchen, and with a sharp knife cut it right through between the centre combs. Spread a piece of paper on a board not less than 15" × 9", and across it lay, at equal distances, two or three narrow tapes, at least 24" long. On the tapes place a comb carefully taken from the skep, arranging the tapes so that they may be passed round the comb and tied at the top. Place a frame over the comb, so that the upper edge of the comb shall meet the top bar of the frame; and, if the comb be too large for the frame, cut it to fit tightly between the top and bottom bars. If the comb be too shallow to fill the frame, put a piece of lath under it, and draw one or more tapes under this. Tie the tapes around the frame and comb; raise the board, frame, and comb together; and set the frame in the new hive. Transfer the other combs similarly, excluding drone comb, and being careful to include all the worker brood in the centre frames **(213)**, and to avoid its being chilled. Close up the frames and dummy: put on the covering and roof: set the hive on the old stand: and run in the driven bees as previously described. Give a little feeding, or uncap some of the honey cells. In a couple of days, when the bees will have fastened the combs in the frames, remove the tapes. Sometimes, wire netting, with a small mesh, is used instead of tapes, and by this means small pieces of comb can be held in position until fastened by the bees. The combs should always hang in the new hive top up, as they were in the skep. The method illustrated here (Fig. 84b) may be made to serve the purpose; several pieces of comb may be held in position in a frame until joined by the bees, or pieces of foundation may be inserted (as in the illustration) either to fill a frame or to substitute worker for drone comb. The laths have ordinary pins driven through them; these hold the pieces of comb, or foundation, and the pins at the ends are driven into the top bar and bottom bar of the frame.

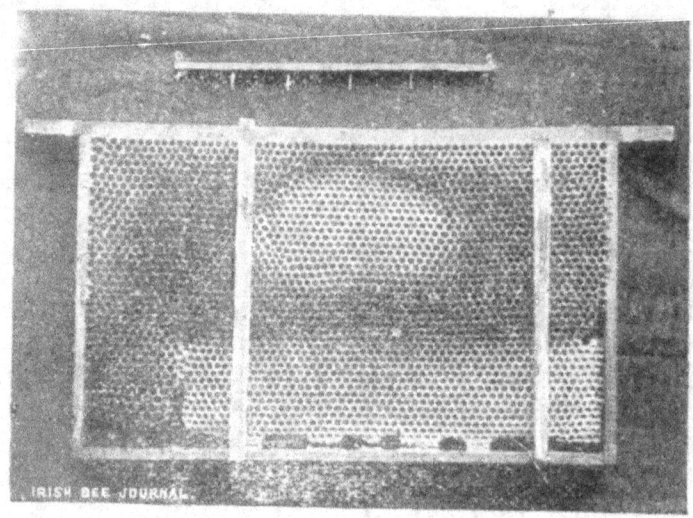

Fig. 84b.
REPAIRED COMB.

254. Automatic Transfer from Skep, or Box, to Modern Hive. —Stimulate the stock **(192, 313, 318)** with the object of having the skep crowded with bees in April, or early in May. When this is attained, prepare a modern hive, as directed **(232)**, cutting a central hole of 6″ or 8″ diameter in the sheet. This hive must now be placed in the position occupied by the skep, and the skep must be set upon the sheet, over the frames. Put on the lift **(87)**, in the summer position (Fig. 116, p. 207). and pack warmly round the skep so that no bees can get out except through the entrance of the lower hive: put on the roof and open the doors. The bees will now leave the hive and return through the regular entrance, passing up and down the frames as they come and go; and as they increase in numbers they will occupy the frames. About ten days after the operation described above, weather permitting, an examination should be made. When brood is found in more than one

CHAPTER XXII.

SURPLUS HONEY.

255. Preparing in Time.—The honey flow **(265)**, in this country, lasts for only a few weeks. To take full advantage of it, and to secure a harvest as large as possible, the bee-keeper should bring his stocks up to their full strength by stimulating breeding **(192)**, and spreading the brood **(193)**, right up to the opening of the flow, and by uniting all weak stocks **(244)**. He should also have for every hive a supply of crates **(103)** fitted with sections of foundation, or drawn comb; or a supply of super boxes **(108)** fitted with frames of wired foundation, or of drawn comb—crates, if he means to work for comb honey; super boxes, if for run, or extracted honey. The crates, or super boxes, should be prepared, wrapped in paper, and laid aside in a safe place well ahead of the opening of the season. It is an expensive habit to defer the preparation of such appliances until they are actually required.

256. Extracted Honey more Profitable than Comb Honey.—Whether he shall work for comb honey, for extracted honey, or for both, each bee-keeper must decide for himself; and his decision should be made sufficiently early to enable him to make his arrangements accordingly. As between the two—section, and extracted honey—the question of profit can be answered only in favour of the latter. The output of extracted honey, where strong stocks are employed, is greater by from 50% to 100% than that of comb honey: the expenses are less, the same combs serving for many years: the marketing is simpler and cheaper, freights being lower, and breakages infrequent: "depreciation" and risks are reduced to a minimum: and the management of the stocks is simplified, there being fewer swarms **(266)**, and, accordingly, less upsetting of the bee-man's arrangements. If there be any extra trouble in dealing with extracted honey; there is less trouble in dealing with the bees. If it be an objection that extracted honey fetches a lower price; there is the compensation that one has about double the quantity to sell, and at a lower cost of production. If the initial cost of an extracting outfit **(134-136)** be a discouragement; the yearly saving in the cost of sections and foundation is a far more than sufficient set off. Suppose the extracting outfit to cost £2 (which is a liberal allowance for a small apiary), the annual charge, at 5%, upon that outlay, with allowance of 15% for depreciation, may be set down at 8s. But the sections and foundation required for four hives may cost £1 per annum; for six hives, £1 10s.; for ten hives, £2 10s., which shows a very substantial economy in favour of working for extracted honey, where more than three strong stocks are employed. (These are not war prices.)

257. Preparing Crates and Sections.—If the crate has been used before, let it be well scalded, washed, and scraped clean from propolis and wax. Fasten a section folding block (which is a piece of wood 4" × 4" × 1½" or 2") to a bench, or table, by a screw through the centre of the block: take a section, and carefully fold it upon the four sides of the block, fastening the ends together: proceed until you have folded twenty-one sections, which will be sufficient for one ordinary crate. If the sections are very dry, and inclined to break at the corners, damp them at the V cuts on both sides before folding.

258. Three Split Sections.—If you are using three split sections **(101)**, foundation can be fixed in three sections at one time. Place three sections in the crate, as shown (Fig. 85), and with the unsplit sides down: between the further side of one of the end sections and the side of the crate, push in a wedge, to hold the sections tightly: draw out the nearer halves of the sections, and drop in a 12¾" × 4" or 4¼" sheet of super foundation: remove the wedge, and, with the follower **(106)** press the sections together, so that the foundation may be gripped. Remove the follower: put in one long, or three short separators **(102)**: add three more sections; and proceed as before until the crate has its full quantity. When the last row of sections is

Fig. 85.
FIXING FOUNDATION IN THREE SPLIT SECTIONS.

in, press the rows together tightly by the follower, and wedge the latter, either by clips, or wooden wedges, to keep all secure. Scrape off the foundation appearing above the sections, and put it aside for the wax extractor **(279)**. By this means sections and crates can be filled very rapidly. If tin, or zinc bars are used to carry the sections **(103)**, the sheets of foundation must be cut, to permit them to drop nearly to the bottom of the sections. Separators between the sections must never be omitted, for, otherwise, the bees may draw out the comb beyond the wood of the sections, making it impossible to pack the latter safely for transit; or, they may build comb to comb, and work ruin in the crate.

259. Split Top Sections.—These sections **(101)** have a bevelled split in one side, to grip the foundation, and, in folding, should have only one half of the split fastened at the dove-tail, and that, what may be called the under-lap half. Place several sheets of super foundation one upon the other, and flush at the ends and sides: on these set the folding block, flush with one end of the parcel; and, with a sharp knife cut through the foundation; proceeding until you have a sufficient number of squares cut, and taking care that the squares will fit properly in the sections. Place an end of one square on the bevel of the section top, and shut down and fasten the other half, fixing the sheet of foundation so that it will hang vertically in the section, and allowing just a little space at the bottom to provide for possible stretching of the square. Place each section, as it is finished, in the crate, with separators between the rows, and the follower and wedges at the back.

260. Unsplit Sections.—Fixing foundation in unsplit sections **(101)** is somewhat more troublesome. The plan frequently recommended is, to prepare a folding block nearly half as thick as the section is wide; the square of foundation is laid upon the block, and the section is placed in position; melted wax is then poured in at the upper edge of the foundation to fasten it to the wood. There are simpler methods which work sufficiently well:—In a saucepan of hot water place a teaspoon, handle down: bend the edge of the foundation at right angles, and place it on the wood so that the square, when fastened, will hang in the centre of the section: with the hot end of the spoon, press the bent edge to the wood: the wax will melt and adhere. In a warm room, the foundation can be readily fixed by pressure. Place the section top-side down, and lay the square of foundation on the inside of the top, projecting about $\frac{1}{4}''$ beyond the centre, and held at the centre by a guide, which may be made from a quarter section cut to the right width. Pressure with any smooth instrument, such as the handle of a dinner knife, will cause the foundation to adhere to the wood. Reverse the section, and arrange the foundation to hang plumb.

261. Preparing Frames.—Frames (97) are generally supplied in the flat, the pieces being made to fit into and grip each other. Fold your frames so that the angles at the corners shall be true right angles, because the frame will not hang properly in the hive or super box if it has been put together out of square. Fasten the frame at the corners with small tacks, or wire nails.

262. Wiring Frames.—All frames (except shallow frames), and especially those that are to come, some day, to the extractor (134) should be wired, so that the foundation may be well supported, and safe from breakage or sagging, with the consequent evils. (118).

"Of course there are some who never wire their frames at all. Happy-go-lucky in their methods, they trust much to the 'lucky'; happy, indeed, until the combs lie broken in the extractor, and then it is not any longer safe for a cat to laugh in the house!"—*J. G. D. in the Irish Bee Journal.*

The frame having been fastened at the corners with tacks, or thin wire nails, bore, with a fine bradawl, two holes through each side of the frame 1½" or 2" from the top bar and bottom bar respectively, at A, B, C, D: half drive two small tacks into the edge of one side close to the holes A and D: pass the wire through A, B and C; draw it over itself between A and B; pass it through D, and give the end a double twist round the tack at D: draw the wire at C, and again at A, until it assumes the position shown (Fig. 86) and is tight enough to "twang": give a double twist round the tack at A:

Fig. 86.
WIRED FRAME.

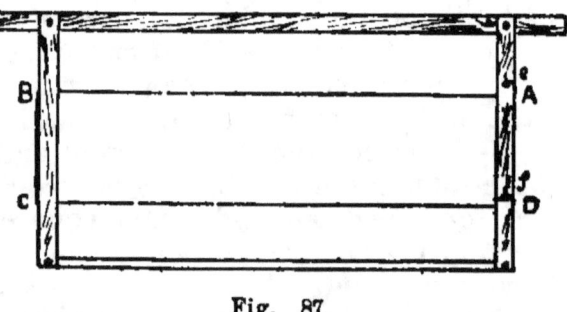

Fig. 87.
WIRED FRAME.

drive home both tacks: and cut off the wire at A. The wire may be fixed in the same way as directed above, but without crossing it, if the parallel system be preferred (Fig. 87). With a little practice, wiring, by either method, can be done very expeditiously.

263. Fixing Foundation in Frames.—The frames, commonly used in this country, have two long grooves cut in the under sides of the top bars; and long, wedge-shaped slips are supplied with the frames. Brood foundation **(112)** is used in frames, and is supplied in sheets of the proper size. To fix the foundation, introduce an edge of the sheet into the groove in the centre of the top bar: place the wedge in the other groove, and press it home, thus gripping the foundation. Place the frame on your wiring board **(118)**, wires uppermost: and, with a heated embedder **(118)**, press the wires into the foundation so that they will grip it, and that the wax melted by the embedder may cover the wire. Put the frames, when completed, in vacant hives, or super boxes; and cover up, safe from damp and dust, mice and moths.

264. Three "Dont's."—Do not allow the wood of the sections to become soiled: dirty sections fetch low prices. Do not fall into the absurd error of using only slips of foundation, or "starters": true economy calls for full sheets in sections and frames **(113)**. Do not put in the foundation wrong side up: bees build their cells with vertical sides, and with angles at top and bottom. **(117)**.

265. The Honey Flow opens at the latter end of May, or in June, according to the district, when nectar is secreted freely in the flowers. When the flow opens, the bees begin to draw out with new, white wax, the cells next the top bars of the frames in the body box. Watch for this infallible sign, and, immediately on perceiving it, give the stock a crate, or super box.

266. Putting on Crates.—If you are working for section honey, bring out a prepared crate **(103)** to the hive; and see that the foundation hangs vertically in the sections, and that separators have not been forgotten. Remove the roof, quilts, and packing, leaving the sheet still on the frames. Take off the riser **(87)**, transferring the porch from it to the body box, if not already done; and set the riser on the ground beside the hive. Give a good coat of vaseline, or petroleum jelly **(174)** to the bottom of the crate, and of the laths on which the sections rest, so that they may not be tightly propolised to the frames by the bees; and set the crate on the edges of the riser. Give a puff or two of smoke to the bees, if you think it necessary,

—though it ought not to be necessary with this operation, and if unnecessary, it should be avoided **(180)**. Roll off the sheet, and draw on the carbolic cloth, as directed **(177)**, to drive the bees down, and to avoid the risk of crushing with the crate any that may be on the frame tops. With a piece of glass, or other scraper, clean off any wax or propolis from the frame tops. If ten or more frames are in the body box, the crate will fit properly across the frames; if less than ten frames are in, the crate must be put on with its sides running with the frames; or, either a piece of wood 17″ long × ¾″ thick, or, extra frames covered with canvas or ticking, must be added behind the dummy, to prevent the escape of bees from under the crate.

PUTTING ON A CRATE.

Take the crate in one hand and hold it just above the carbolic cloth (without touching the latter, lest the vaseline be rubbed off), and in the position which it is to occupy, as illustrated: with the other hand draw out the carbolic cloth, quickly setting the crate upon the frames, and arrange the crate to fit so that no bees may escape outside it. Put on the sheet and the riser: pack all round the crate with warm stuff, or newspapers: spread a couple of newspapers on the sheet: and add the quilts and roof. If crates be not kept warm, the bees will be slow to take to them, and their work in the sections will be indifferently performed. For the proper filling of sections, it is necessary that the bees should be well crowded into the crates; and this points the difficulty of working for perfect sections and at the same time restraining swarming; for, crowding, as we know, is an incitement to swarming, and without a certain amount of crowding the sections are likely to be built with pop holes at the corners, and to be imperfectly drawn and fastened next the wood. Bees are sometimes slow to take to the first crates at the beginning of the season. They may be encouraged to start work above if some sections with drawn out comb, or with comb containing a little honey be given them. For this purpose it is useful to keep over, in a warm, clean place, some unfinished sections from the previous year **(278).** The Hanging Crate **(107)** may be used to get sections started in the brood chamber, and these, with the adhering bees, may be inserted in the first crates; when, if the latter be kept warm, work will be commenced there.

267. Putting on Super Boxes.—Super boxes **(108),** with frames for extracting, are vaselined, and put on in the same way as are crates, with these variations:—(1), excluder zinc is generally used **(109),** and (2), as super boxes with frames are heavier than crates of sections, the manner of removing the carbolic cloth and putting on the super box, so as not to allow bees to fly up during the operation, must be modified. If you have an assistant, get him to pull off the carbolic cloth while you hold the super box in position just above the cloth, ready to set it on the frames. If you are alone, stand behind the hive: hold the super box down to the cloth, which you will catch by the edge with your right hand, as illustrated: give your hands a quick jerk to the right, pulling off the cloth; and at once set on the super box. With a little practice, this can be done so rapidly that, the bees having been driven down by the carbolic, not one will have time to escape before you have the super box on, and covered. Another simple method is to stand at the side, and slide the super box along the frames, pushing the carbolic cloth before it. The objections to this method are, that a little propolis or wax on the frame tops may cause a difficulty in sliding on the super box; and, that unless your hive is so constructed that the ends of the top bars are held by the outer hive walls, as they should be **(86),** the sliding on of the super box may press the frame ends against the inner walls of the hive and may crush many bees, perhaps even killing the queen.

PUTTING ON A SUPER BOX.

268. Use of Excluders.—Frames in the brood chamber should hang one and a half inches from centre to centre. But in the super box, spaces of two inches from centre to centre give better results, because they enable the bees to build longer cells, which, of course, hold more honey, and are easier to uncap for extracting (276). When frames are used in this way, the spaces between the shoulders must be filled, to prevent the escape of bees. It is also claimed, for the two-inch spacing, that the queen will not deposit eggs in combs so spaced because of the depth of the cells, and that, therefore, excluders under the frames are unnecessary. Many experienced beekeepers work their supers in this way, and avoid what, certainly, is an objection to the excluder, namely, the impediment it offers to bees loaded with honey (109). But, it is better to use an excluder than to have the combs, intended for honey, occupied by brood; and, when frames in supers are spaced one and a half inch from centre to centre, as in the brood chamber, excluders should always be used. With respect to the use of excluders under crates of sections, opinions differ widely. In some districts, and with some stocks, excluders are found to be necessary; in other cases, not. In cold, wet seasons, queens will often go up to the warmer part of the hive and take possession of the crates; and bees have been known, in exceptionally unfavourable seasons, to rear queens in the sections, and even to swarm, leaving frames of foundation in the brood chamber untouched, and combs unoccupied. But, making due allowance for the vagaries both of the climate and the bees, crates may be generally used without excluders underneath if sufficient room be given to the queen in the brood chamber (193). When excluder zinc is used, it should lie flat upon the frames, leaving no space at the edges for the queen to ascend. It may be laid upon the carbolic cloth, and held while the cloth is drawn from under it; then, if the cloth be spread for a moment upon the excluder, the super box, or crate can be put on as directed above.

269. Tiering Crates.—In a good season, a strong stock may require a second crate within a week. If honey is coming in rapidly, and the days are fine, the second crate may be given when it is seen that the bees have drawn out the foundation in their sections, and are storing honey there. In the height of the honey flow, swarming may be provoked by a day's delay in giving more super room when it is required. Give the second crate underneath the first one. Prepare it as before (257); and set it on the riser beside the hive. Subdue the bees with smoke. If the first crate has been well vaselined, it will come off easily; if not, prise it up at the corners, and insert bits of broken sections there. If the laths on which the sections rest are too thick or too thin, or if they have sagged, the bees will, probably, have fastened them and the sections to the frames (103); and the loosening may exasperate both the bees and their owner. Grasp the crate with both hands, and twist it gently to right and left until it is loose for removing; then twist it back to its original position for a moment. If it is too tightly fastened to be loosened by twisting, draw a piece of wire under it to cut the connections, and prise it up taking care not to allow bees to escape. If you have an assistant, stand at the back of the hive. Raise the crate just free from the frames, and take it off along, and not across the frames; your helper following it closely with the carbolic cloth, as illustrated (Page 151). If you are alone, stand at the side of the hive. Take the carbolic cloth at two corners between the fingers, with both hands, and let it hang down outside the hive as shown—at the side, if your frames hang parallel with the entrance; at the front if the frames run from

REMOVING A CRATE.

front to back—in which case your position will be at the back. Grasp the crate, ease it, and take it along, and not across the frames, letting the carbolic cloth cover the frames as you remove the crate. Hold the crate for a moment over the cloth, to cause the bees to run up into the sections; then, set it upon the second crate, and lift both back on to the frames, drawing away the cloth as directed above (267), and settling the crates evenly upon the frames. The sheet and quilts not having been taken off the first crate, the operation may be carried out without allowing any bees to give trouble. This tiering up of crates (Fig. 88) may be continued while the honey flow lasts, the empty crate being placed underneath. The upper crates may be removed when finished, and before the faces of the combs become soiled by the constant coming and going of the bees: but, if the crates be tiered up until the close of the honey flow, the honey will keep its flavour best on the hive; and the bees, having so much room, will be less inclined to swarm, and, not being deprived of their stores, will be less inclined to give trouble. In a good season, as many as five crates may be required for a strong stock; in which case, if tiering be practised, an extra make-shift riser will be required, and assistance in lifting the crates together will be necessary. It is a good plan to use a Divisional Crate **(104)** for the last addition to the tier, and at the close of the season where tiering is not practised; because, seven or fourteen sections may be given, when the season has advanced too far to admit of twenty-one sections being added with any prospect of their being filled and sealed; and, because the parts may be removed as the sections in each are completed, the last, unfinished sections being placed over the centre of the cluster. The Hanging Crate **(107)** may be used to get sections completed in the brood chamber.

Fig. 88.
CRATES TIERED

270. Doubling and Storifying.—When extracted honey is being worked for, "Doubling" may be practised with excellent results, both as regards the harvest that may be obtained from it, and the restraint it exercises upon the swarming impulse. About three weeks before the opening of the honey flow **(265)**, take, from a strong stock, all the frames containing brood, except one, on which the queen must be left: return the adhering bees to their hive; and fill the vacancies with frames of comb or of foundation. Place the frames of brood in another body box, or in a super box, and set them on top of a second strong stock, with an excluder **(109)** underneath, thus doubling the hive (Fig. 89). The stock, increasing daily by the emerging brood in both storeys, will become very strong, and will be capable of storing honey very rapidly in the upper frames as

SURPLUS HONEY. 149

Fig 89.
HIVE DOUBLED.

the brood there hatches out. The combs of honey may be removed from above to have their contents extracted, and to be returned at once to the hive for refilling, drones being removed; or, the two storeys may be used as brood chambers, and a third and a fourth storey may be placed on top for honey only, the excluder zinc being placed above the brood chambers, to safeguard the upper storeys from the queen's attentions. When the combs in the uppermost storey have been filled, they can have their honey extracted, and may be returned upon the excluder, the unfinished storeys being placed above them.

Fig. 90.
SKEP WITH SUPER CASE AND ROOF.

271. Supering Skeps.—Skeps with flat tops (**75**) may be supplied with crates. A "riser," or case (Fig. 90), 9" deep, and large enough to hold a crate, is fitted with a false bottom 4½" from the top. A hole in the false bottom, corresponding with the hole in the top of the skep, is covered with excluder zinc. The riser is fastened to the skep by four nails; and a deep roof permits the use of two tiered crates.

272. Removing Supers.—When the honey flow is over, and the nights grow chill, the bees will begin to take down honey from the supers to the brood frames. Therefore, supers should be removed in good time, and it is better to remove them a little too soon than a little too late. This is an operation which

T. W. H. BANFIELD SUPERING A SKEP.

requires some care, in order to avoid the risk of setting up robbing (**307**), and of exasperating the bees. The point to be aimed at is to take away the supers so skilfully as neither to expose honey to the bees outside, nor to put too severe a strain upon the patience of the bees within.

273. Use of Cone Escapes.—On fine, warm days, the cone escapes (Fig. 91) in the hive roof may be used, with some success, to clear the supers of bees. Through a round hole 1½" in diameter, in the front gable of the roof, a cone escape is passed from the inside, and tacked; and a second cone is

SURPLUS HONEY.

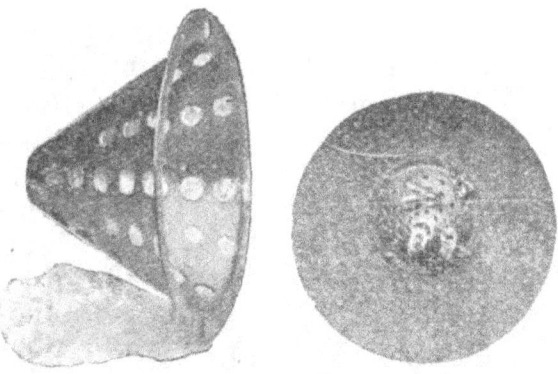

Fig. 91.—(a) CONE ESCAPE.
(b) DANGER! QUEEN ON THE CONE.

fitted on the outside (88). Early in the day, having lifted the crates or super boxes, as directed above, draw, or get an assistant to draw, a towel over the frames, covering them completely. Set the supers back upon the towel: remove the sheet and other coverings: and put on the roof. If there be no hole in the towel, the bees in the supers will be unable to get down to the frames: they will come to the top of the supers, and, seeing the light through the cones, they will pass out that way, returning to the hive through the usual entrance; and in the evening the supers should be free from bees. If there be a ventilator in any other part of the roof, it should be closed, so that the bees may see light only through the cones. There are, however, some objections to the use of cone escapes for this purpose; for, if the day be cold or wet, the bees will not leave the hive roof; and if there should be, in the crates, young bees that have not yet been on the wing, they may be lost on emerging from the cones.

REMOVING A FRAME SUPER.

274. The Super Clearer (Fig. 92) has simplified the once laborious and trying operation of removing surplus honey from bees. It is useful, also, when extracted combs require to be cleaned up by the bees before being stored away (278). It consists of a 2″ × 2″ frame of wood, 17" long × 15¾" wide, in which is a panel ¾" thick, bee space being thus provided on both sides of the panel when the clearer is in position on the hive. In the centre of the panel is inserted a bee escape, and at one side, near the frame, is a 1½" hole, which may be opened or closed by a shutter worked from the edge of the frame. When the clearer is placed on a hive, and supers are set upon it, the bees pass down from the supers through the escape, and cannot return. The side hole is opened only when it is desired to admit the bees to the combs for cleaning up purposes **(278)**. The combs are placed upon the clearer, and the side hole being open, the bees quickly take down every particle of honey, and leave the combs perfectly dry. The shutter is then closed, and the bees clear through the escape. The super clearer has this advantage over the cone escape **(273)** that, whereas the cone escape operates only during daylight, and upon genial days, the super clearer can be worked by day and night, and no matter what the weather outside may be. The "Porter" Bee Escape (Fig. 93), for super clearers, is a metal box, with an arrangement of delicate springs which permit bees to pass out. Sometimes, however, an "awkward drone," getting stuck in the passage, bars the way against all others, and thus renders the escape inoperative. In use, the round hole is on top. The "Federation" Bee Escape

Fig. 92.
SUPER CLEARER.

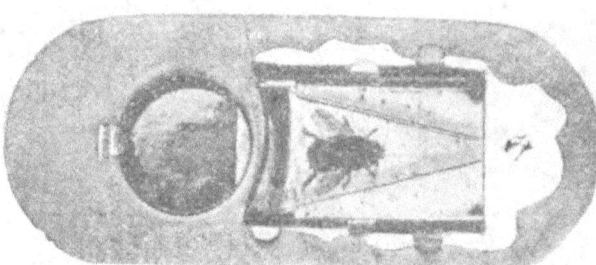

Fig. 93.
PORTER ESCAPE.

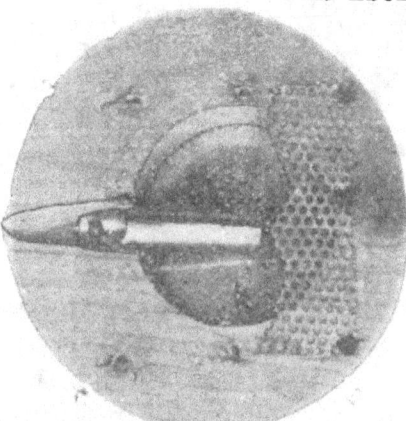

Fig. 94.
"FEDERATION" ESCAPE.

(Fig. 94), for super clearers, is not so liable to get blocked as is the Porter escape. The bees pass through a tube which is large enough to admit drones freely; they then drop upon a tin platform, and get down through a hole in the centre, under the tube. In the illustration, the perforated zinc has been cut, in order to disclose the tube. In use, the perforated zinc is on top. The escape is made of timber $16\frac{1}{2}'' \times 14\frac{1}{2}'' \times \frac{3}{4}''$. On this a frame of $1\frac{1}{2}'' \times \frac{1}{2}''$ stuff is nailed above and below, projecting $\frac{1}{2}''$ all round. Thus the escape measures $17\frac{1}{2}'' \times 15\frac{1}{2}''$ out to out. The hole in the panel is $2''$ in diameter; the tube is $2''$ long and $\frac{3}{4}''$ in diameter; and the hole in the piece of tin underneath is $\frac{1}{2}'$. A groove is cut in the panel to admit the tube. A piece is cut from the upper side of the tube, where it extends outside the perforated zinc, to give free access to the bees. At one side of the panel, and $\frac{1}{2}''$ from the inside edge of the frame, a hole $1\frac{1}{2}'' \times \frac{1}{2}''$ is cut through. When not required, this hole is closed by a tin slide $3\frac{1}{4}'' \times 2'$, which is slipped under the frame, and turned up outside; a slit is made in the slide, and a nail driven through the frame and slit permits the slide to move in and out, *i.e.*, to cover the hole when supers are being cleared, or to leave it free to the bees when the object is to return combs for cleaning up purposes.

275. Use of the Super Clearer.—Super clearers permit the bees to pass down from the supers without having to leave the warmth of the hive, and, therefore, they may be used by day or night, and in all kinds of weather. Examine the clearer, and see that the escape is in working order, and that the side hole (274) is closed. Set the clearer beside the hive, right side up: lift the supers; set them on the clearer; and put all back on the frames, without removing the sheet and quilts. The bees will pass down to the body box, and, next day, the supers will, probably, be found emptied of bees, and may be removed, little disturbance of the colony having been caused by the operation (397).

CHAPTER XXIII.

EXTRACTING HONEY.

276. Extracting.—Let your extracting be done in-doors, so as not to provoke robbing. Where this is not always possible, extracting may sometimes be safely done out of doors in the height of the honey flow, when bees are not inclined to rob, and if due precautions be taken to protect the combs and honey from any little pilferer who, having once got a taste of the sweets, may quickly lead many others to the spot, and thus set up general robbing at the hives **(307).** When there are many combs to be extracted, time will be saved if **two** uncapping knives be used **(135).** See that the edges are as sharp as they can be made, and heat the blades in a vessel of hot water while in use. Lay a strong lath across a crock, or other similar vessel. Take up one of the frames; hold it with one end resting on the lath; and, with your uncapping knife, working from the bottom upwards, pare off the cappings, inclining the frame towards the knife so that the cappings, as they are pared off, may fall into the vessel underneath. Having uncapped both sides, proceed in the same way with a second frame. Put the frames, ends up, in the extractor **(134),** one in each cage, and so that, in revolving, the bottom bars of the frames will travel first. Turn the handle, slowly at first, and increasing in speed by degrees, being careful not to revolve the cages so furiously as to break the combs. The centrifugal force will throw the honey out of the cells. When one side of each comb is finished, or nearly so, reverse the frames in the cages, and extract from the other sides. With new, soft comb, it is better to extract only about half from one side, then reversing, and returning to the first side to finish. Honey, being more fluid when warm than when cold, may be extracted more easily if it has just been taken off the hive, and if the extractor is warmed with boiling water immediately before use, and is kept near a good fire during the operation. With extractors which are not geared, the turning of the handle is often a laborious job when a number of combs have to be done. The work may be simplified by arranging what may be called a "cord gearing" by means of a strong cord, about four feet long, with a loop on one end, which is

EXTRACTING HONEY.   [J. G. Digges

slipped over the handle. The operator then stands out from the extractor (as illustrated); gets the cord at right angles with the handle-crank; gives a slight pull at the cord with one finger, thus revolving the cages, and assisting each revolution by another slight pull. With a few minutes' practice one can extract honey in this way with little exertion. Sections that are not completely filled, or that are otherwise unfit for the market, may have their honey extracted in the same way, six sections being placed in each cage. All cappings may have the adhering honey pressed out, and may then be rendered into wax (279). Heather honey requires special treatment, and can be thoroughly separated from the comb only by means of the honey press (137). A piece of clean cheese cloth, large enough to fold down to the bottom, is placed over the top of the press. The combs are placed in the press vertically, as they hung in the hive; the cheese cloth is folded over them; the plunger is then brought over, and the screw is revolved. The honey falls to the drawer beneath, and the wax is lifted out in the cloth. (See also 400-402.) A potato masher will press small quantities.

277. Straining and Ripening.—Extracted honey may be strained from the extractor into a ripener (136), and should be left, for a few days, covered, in some place with a high temperature, after which the honey may be bottled for market (306), the thin honey on the top being used for bee food only.

278. Cleaning Extracted Combs.—Frames and sections, having had their honey extracted, may be given to the bees to clean up before being stored away for use in the next season. Place a super clearer (274) on the frames of a strong stock; and, in the evening, when the bees have ceased flying, draw open the side trap; set the frames and sections on the clearer; and cover up, safe from marauding bees. The bees of the colony will come up through the side trap, and carry down all the honey, leaving the combs clean and dry. The side trap may then be closed, when, the bees above the clearer will pass down through the trap in the centre, and will not be able to return. The same object can be attained by giving the combs behind the dummy, leaving a bee space between the dummy and the floor board; or behind the "Federation" dummy (95), with excluder zinc attachment. Frames and sections, when cleaned, should be removed, wrapped carefully in clean paper, and stored away in some dry place, safe from mice, flies, and other adventurers (371). It is a good plan to keep over, for the following season, a few unfinished sections, having a little honey in them, for use as "bait sections" in the first crates; thus coaxing the bees to take to those crates, and to begin work at once. (266).

CHAPTER XXIV.

EXTRACTING WAX.

279. Use of Wax Extractors.—Wax is so valuable that no careful bee-keeper will permit the smallest piece of it to be wasted. Clippings of foundation, cappings removed from combs, and old or broken combs, should be collected and rendered, either for sale, or for manufacture into foundation. The light coloured, and the dark coloured wax should be rendered separately, as the former fetches the higher price. As already described, both the Solar Wax Extractor **(139)** and the Steam Wax Extractor **(140)** give good results. The combs should be soaked for twenty-four hours in cold water before being rendered. The best coloured wax is obtained by means of extractors; but a large amount of wax remains in the debris, and, it is only by subjecting it to considerable pressure while hot, that the mass can be made to yield nearly all its wax. This is especially the case when old combs are being dealt with. (See also **404**.)

280. Extracting by Boiling.—A third method of rendering wax may be adopted as follows:—Soak the wax in water for twenty-four hours, as directed **(279)**. Into a canvas bag, or a clean, closely-made sack, put a large stone: throw in also all the wax that is to be melted: and tie the bag tightly. Place the bag in a farm boiler, or a large pot, of rain water, with a piece of wood under the bag to prevent burning. When the water has boiled for a couple of hours (or less in the case of clean, fresh combs), let it cool: and, when cold, remove the cake of wax from the top. Scrape the dirty wax from the bottom of the cake into the bag, and boil it again for two or three hours, when, on cooling, a cake of inferior wax may be taken off. The first cake should be broken up, and put into an enamelled vessel of hot water, and the vessel set in a pot of boiling water near the fire until the wax melts, after which it can be poured into shapes, and cooled slowly, as before. The colour of the inferior wax may be improved by adding a little vitriol to the water in which it is boiled, in the proportion of three tablespoonfuls of vitriol to one gallon of water. Smaller quantities may be wrapped in a piece of cheese cloth and suspended over a vessel of water in the oven. When melted, cooling must be very gradual to avoid cracks. The wax cake may be removed from the vessel of water when cold **(404)**.

CHAPTER XXV.

QUEEN REARING AND INTRODUCTION.

281. Old Queens.—Attention has already been called to the necessity for supplying young, prolific queens to all stocks requiring them **(212)**. Too much emphasis cannot be laid upon the fact that queens past their second year are past their prime **(188)**. The bee-keeper who desires to work his stocks to the best advantage, will not fail to supplant all such queens. He will not be content to leave this most important part of his work to take care of itself. Either he will purchase good queens from other queen-raisers, and thus introduce new blood into his apiary; or he will do his own queen rearing. It is quite certain that, in this country, the supplanting of old queens is not attended to as it should be. This may be due to the supposed difficulty of rearing queens. There are so many capable workers of bees who are satisfied with average harvests, satisfied with the second best, and deterred from attempting to rear their own queens, because of the trouble or the difficulty which they think that that part of a bee-man's work involves. Bee-keeping can never be raised to the level to which it ought to attain, until queen-rearing is practised, not by the few, but by the many.

282. Defective Queens.—Sometimes queens are found to be defective, and their places must be supplied by fertile queens if the colony is to be preserved. Queens that have not been impregnated within three weeks after leaving the cell usually become drone breeders **(188)**. Queens that have been chilled, or half starved, or that have been "balled" **(296)**, or injured in the hive, may lose their fertility and become useless. The bees will generally supplant such queens, but they cannot do so unless the conditions are favourable **(196)**, nor without loss of valuable time.

283. Queenlessness.—Beside the necessity for supplanting aged, and defective queens, there often arises a necessity for supplying fertile queens to stocks whose queens have been lost, or killed, or that have died natural deaths. This is an urgent need which, whether it be observed or not, presents itself more frequently than many suppose. At the time of swarming, queens are sometimes lost if they alight apart from the swarm, undiscovered by the bees or the owner. Unskilful, or careless manipulations of frames are accountable for the crushing and death of many queens **(182)**. When they leave the hive to meet the drones, some queens, either through some defect of their wings hindering their return, or through the assaults of birds, or of strong winds, fail to reach their homes again. By far the largest number of lost queens become lost through their inability to recognise their own hives, when returning from their wedding flight **(147)**. And this disaster is frequently due to the habit of using hives so close together, and so similar in their make, colour, and situation, that, although the virgin queen takes all due precautions to mark the position of her own hive before her flight **(21)**, it is next to impossible for her to distinguish it from the others when she returns, and, entering a strange hive by mistake, she is immediately killed. This is a fact of sufficient importance to point the necessity for keeping careful watch over all casts and swarmed stocks until one is satisfied of the mating and laying of the queens; and, also, to lead to the re-arrangement of any apiary **(391)** in which the conditions are such as favour the loss of newly-mated queens. And it should be noted that, although a colony deprived of its queen can, in certain circumstances, supply the loss **(17)**, if the loss occur when there are neither eggs nor larvæ under three days old in the combs (as in the case of a swarmed stock, or a cast), a new queen cannot be raised, and the colony, if left to itself, must dwindle and perish.

284. Signs of Queenlessness.—When a colony has become queenless, the fact may soon be discovered by observing the conduct of the bees. They hurry about the hive, in and out, and over the porch, sides, and roof, as if in search of their lost mother. This may continue for two or three days; after which work is resumed, but, in a listless, half-hearted way; the bees returning from the fields loiter about the alighting board, with little apparent anxiety to enter the hive, and a general air of indifference prevails in the colony. In spring, they carry in little or no pollen, there being no brood to feed. In late autumn and winter, they permit the drones to remain in the hive. Such signs as these will indicate to the owner that something is wrong with the colony; and, if on examining the frames he finds no queen, and neither eggs nor brood, at a time when they ought to be present, or only the eggs or brood of a drone-breeder **(188.200)**, he will know that he can save the colony only by taking measures for re-queening it, or by uniting it with another stock.

285. Nucleus Hives.—The proper time to begin preparations for queen rearing is in the winter, when, a supply of nucleus hives, sufficient to meet the needs of the apiary, should be prepared. Hives which are not required for other purposes, can be temporarily transformed into nucleus hives, so that they may be turned to use again for swarms and stocks at a moment's notice. Divide the hive into three parts, by inserting two close-fitting dummies **(93)**. Make an opening, ⅜″ deep **(371)**, in the back, and another midway in one side, level with the floor board, to form two additional entrances. Part of a broken section, tacked to a piece of inch wood nailed or screwed below the augur hole, will serve as an alighting board; and a porch, or rain shoot, may be similarly constructed. When the hive is required again for a stock or swarm, all that will be necessary will be to remove one or both dummies, and to stop the augur holes with corks. Nucleus hives may be inexpensively made up from grocers boxes (Fig. 95), provided that the timber be sweet and clean.

Fig. 95.
MAKESHIFT NUCLEUS HIVE.

They should be made to take three, four, or five frames, and should measure internally 14¾″ long × 9″ deep. If the sides be made 17″ long, and if the end pieces be 8½″ deep and be nailed 14¾″ apart, two pieces can be fastened to enclose the frame shoulders, and the bottom board can be also 17″ long, to provide an alighting place for the bees. The sides may be made of 11″ timber, which will leave a space of 2″ above the frames for quilts, etc.; but a shallow riser **(87)**, which would admit of the use of a feeder, would be preferable. A piece of board, two or three inches longer, and wider than the hive, may be set on for a roof, and if a brick or a heavy stone be laid on top, it will keep all secure. Legs may be added, or the hive may be set upon a couple of bricks, with a tilt to the front, or back, to throw off rain.

286. Queen Rearing.—Early in the spring, the scene of operations will be transferred to the stocks which are to be used. It should be borne in mind that, to secure the best results, the young queens should be reared when the stocks are strong, when nectar is coming in rapidly, when drones are on the wing, and when the condition of the stocks is such

QUEEN REARING AND INTRODUCTION. 161

as prevails in the swarming season; and, also, that the queen rearing should be from the eggs of those queens which are in their prime, *i.e.*, in their second year, and which have distinguished themselves as the best in the apiary by reason of the excellence of their laying powers, and the vigour and diligence of their progeny. It is desirable, further, that the mating of the young queens should be with the best drones, the temperament of the progeny being largely influenced by the male element. If you have a sufficient number of stocks to permit of two being set apart for the purpose, select two of the most desirable (A and B), and keep up regular stimulative feeding **(192)**, and the other methods already described **(193)** to bring the two stocks rapidly to full strength. When the hive B (which is to rear the drones), is sufficiently strong, insert two drone combs, or two frames of drone foundation, in the centre of the brood nest, and do not permit the feeding to flag, so that drones may be flying from that hive in time to fertilise the young queens. To carry the preparations further, drone breeding may be limited, or prevented in the other stocks, by cutting out or removing all drone comb, and by supplying only worker comb or worker foundation. **(195)**.

287. Using a Swarmed Stock.—Suppose that the good stock (A) sends off a prime swarm. The swarm may be hived, and placed upon the stand previously occupied by the parent stock, and may receive from the latter the supers, and the flying bees returning to their old stand **(240)**. That swarm should give a good account of itself. The parent stock is removed to another part of the apiary and examined. It will be found to have a number of queen cells, and a good supply of young bees upon, let us suppose, nine or ten frames. Now, a prepared hive, with three compartments, or three nucleus hives, being at hand, the combs with the bees from the parent stock are inserted, so as to form three nuclei, each having one or two queen cells. When a queen has been hatched and fertilised, she can be introduced to a stock which requires requeening, and the nucleus from which she has been taken may be used to rear more queens. Eventually the two dummies may be removed, and the bees may be united into one stock **(246)**, or, the bees and frames may be given to other stocks. This operation, it will be seen, has the recommendation of extreme simplicity.

288. The Returned Swarm Method.—A plan which is sometimes adopted is, to let the best stock swarm, and from this, the prime or first swarm, to remove the old queen, allowing the swarm to return to the hive. Nine days later the swarm, increased in size, will re-issue, headed by a virgin queen.

The swarm is then hived on the stand of the parent stock, which latter is moved to a new stand, or is divided into four or five nuclei, each provided with a ripe queen cell of its own rearing.

289. Using an Unswarmed Stock.—When the desired drones begin to hatch out in hive B **(286)** insert a frame of worker comb, or of worker foundation in the centre of the brood nest of hive A. On the third or fourth day, if eggs have been deposited in that frame, transfer the queen and three frames, one of brood and two of honey, with the adhering bees, to a nucleus hive, pushing a little grass into the entrance to prevent the bees from returning at once to their old home. Supply syrup if necessary. Remove also from hive A all combs having unsealed larvæ, returning the adhering bees, and give the combs to other stocks. Now take out the frame which you inserted in the centre of the brood nest, and in which the queen has deposited eggs, and with a penknife, cut "scollops," or V-cuts, from the bottom of the comb up to where the eggs are found, and with a pencil or a match, enlarge the cells at the apex of each scollop, to encourage the bees to build a queen cell there: or, cut holes through the comb immediately under the eggs, returning the frame as quickly as possible, and covering up the brood nest warmly **(338)**. If honey and pollen are not coming in plentifully, you must supply them artificially **(192)** during the next few days. Nine or ten days after the scolloping of the comb, there should be a quantity of queen cells upon it; and you must then prepare nucleus colonies to receive them.

290. Forming Nuclei.—Take, from a strong stock, one frame of honey and two frames of brood, with the adhering bees, supplying their places with frames of comb or of foundation, and insert the removed frames and bees in a nucleus hive **(285)**, taking care to leave the queen in the parent hive. Stop the entrance of the nucleus hive with grass: arrange obstacles about it as directed elsewhere **(156)** to cause the bees, when they fly, to mark the new situation: supply food: and, should the colony become reduced too much by bees returning to the old hive, shake some more, and preferably young bees, into it from the parent stock, or from other stocks, using, in the latter case, the precautions described under the head of "Uniting Bees" **(244)**. Place the nucleus at some distance from the other stocks, and continue the operation until a sufficient number of nuclei have been formed.

291. Inserting Queen Cells.—Having, on the ninth or tenth day, formed your nuclei, supply them with ripe queen cells on the following day, by which time they will have realised their queenless condition, and will be prepared to receive assistance. When queen cells are ripe, that is, within two or three days of hatching, the bees remove some of the wax from the points of the cells, thus roughening them, and facilitating the egress of the young queens, and enabling the bee-keeper to recognise the cells as ripe, and ready for use. You must remember that frames with queen cells will not admit of being cleared (184) by shaking or thumping, and that on no account must the royal brood be suffered to become chilled during the operation of transfer. Gently drive the bees off one of the ripe cells with a carbolic feather: cut out the cell with a piece of the comb above it (Fig. 96): return the frame to its hive: and insert the queen cell between two combs of a nucleus, fastening it by thinning the attached piece of comb and turning it down upon the frame-top, pressing it flat. If the queen cell has been built upon the face of a comb, cut round the queen cell, right through the comb; and, from the brood comb of the

Fig. 96.
QUEEN CELL, CUT OUT FOR INSERTION.

nucleus cut out a piece the same size; and insert in its place the piece with the queen cell. The cell must not be pressed in the least by the fingers. A couple of days later, examine to see whether the cell has been accepted; and, if it is found to have been destroyed and other cells to have been built, remove the latter, and give another ripe queen cell. The risk of chilling the queen brood may be avoided by heating in the fire a 4-oz. weight, or other piece of metal, until it is as hot as you can bear in the hand: place this in a small box, and cover it with three or four thicknesses of felt: lay the queen cells upon the felt, close the box and put it in your pocket. This will keep the cells and their brood warm while you are preparing to insert them in their new positions.

292. Management of Nuclei.—When all the nuclei have been supplied with queen cells, they must be warmly covered up, and gently fed with syrup. There is a danger of the bees of a nucleus leaving the hive with the young queen when the latter takes her mating flight. If a frame of young brood be given to them, they will not be likely to forsake it. The original queen of hive A (289), with her attendant bees, may then be returned to her old home; or she may be used elsewhere, in which case the parent stock should have one or two of the queen cells left to it. If, however, more queens be required, the parent stock may have another frame of eggs from the same queen given to it, and that stock may be kept at queen rearing all the season. When the young queens of the nuclei have been fertilised, and have begun to lay, they may be introduced **(295)** to the stocks which need them, or may be utilised otherwise as desired.

Photo from life]   Fig. 97.   *[by J. G. Digges.*
QUEEN REARING—QUEEN CELL ON PREPARED FRAME.

293. Using Two Stocks.—When two stocks can be spared, one to produce the eggs, and another to rear the queens, the following plan may be adopted with good results:—Select the best stock (A.), and the second best (C.). Into the centre of the brood nest of A. put a frame of foundation. From C. remove the queen and three frames, one of brood and two of honey, with the adhering bees, and place them in a third hive, taking the precautions suggested above **(289)**. Three or four days later, open hive C. and rub off all queen cells that have been formed upon the combs. Take from A. the frame given it, which should have a quantity of eggs of the right age in its cells; "scollop," or otherwise prepare it as directed above **(289);** put it in the middle of the brood nest of C., and leave it for ten days, when you should find upon it a quantity of queen cells within two days of hatching. Now form nuclei, as described above **(290),** and give to each one or two queen cells, and one or two queen cells to the nucleus in which is the queen of stock C., which queen you may return to C. by the "direct method" to be described below (299). By this means you have queens raised from eggs laid by your best queen, and nursed by the bees of another good stock, which is always desirable. The illustration above (Fig. 97) shows a modification of the former plan, which has some distinct advantages. Cut two pieces of wood, 3″ × ⅛″, and long enough to fit into a frame. Make two saw-cuts in an edge of each, 2½″ and 5″ respectively from one end: tack them into the ends of the frame, with the saw-cuts to the centre: cut two thin ⅛″ laths to slide in and out of the saw-cuts, thus making three miniature frames. Remove the queen from stock C. Three days later cut a 4″ piece of comb with eggs from your best stock A. Make strips of this piece, by running a knife through alternate rows of cells: with a sharp, hot knife cut down the cells on one side to half their depth: destroy every alternate egg on that side with a match: fasten the strips (prepared cells downwards) to the top bar and laths with melted wax: and give the frame to stock C., after rubbing off all queen cells started there. Nine days afterwards you should have a number of queen cells built on the top bar and moveable laths, and these cells you can distribute as required. Give more strips of comb with eggs to stock C. You can keep that stock rearing queens all the season. The illustration above, which is from a photograph, shows the prepared frame with all the queen cells removed, save one left to the bees to enable them to re-queen themselves.

294. Distributing the Nuclei.—When the nuclei are no longer required for queen rearing, the bees and frames can be distributed among the stocks in the apiary, or they can be formed into one stock, headed by a young queen.

295. Queen Introduction.—Most of the methods of safe introduction at present in use are based upon the belief, gathered from experience, that, if a colony be really queenless, and if a new queen can be introduced, and protected from assault until she has acquired the peculiar scent of the colony, and until the bees have become accustomed to her, she will be accepted. The operation requires care on the part of the bee-keeper, because, there is always some danger, and often much danger that the queen may be roughly treated and even killed.

296. Balling the Queen.—When the bees of a colony are intent upon regicide, they usually surround the queen, enclosing her in a living ball, so firm and close that it is not always easy to break it up. This is known among bee-keepers as "balling the queen." A strange queen, carelessly introduced, or liberated in a colony that is being attacked by robber bees, so that the queen may be mistaken for an enemy; and even the queen of the colony, when manipulations are carried on at unseasonable times, may be balled and hugged to death, before the owner can discover the mischief and remedy it. The poetic fancy of Maeterlinck who, while he admits that "bees are not sentimental," will not allow the possibility of individual disloyalty in the hive, attributes the balling of the queen to a law which "invests her person, whoever she be, with a sort of inviolability," and prohibits the direct assault of any one bee—

"No bee, it would seem, dare take on itself the horror of direct and bloody regicide. Whenever, therefore, the good order and prosperity of the republic appear to demand that a queen shall die, they endeavour to give her death some semblance of natural disease, and by infinite subdivision of the crime, to render it almost anonymous. They will, therefore, to use the picturesque expression of the apiarist, 'ball' the queenly intruder; in other words, they will entirely surround her with their innumerable, interlaced bodies. They will thus form a sort of living prison, wherein the captive is unable to move; and in this prison they will keep her for twenty-four hours, if need be, till the victim die of suffocation or hunger."—*Maeterlinck*.

Huber thus describes the balling of the queen—

"If another queen is introduced into the hive within twelve hours after the removal of the reigning one, they surround, seize, and keep her a very long time captive, in an impenetrable cluster, and she commonly dies either from hunger or want of air. If eighteen hours elapse before the substitution of a stranger-queen, she is treated, at first, in the same way, but the bees leave her sooner, nor is the surrounding cluster so close; they gradually disperse, and the queen is at last liberated; she moves languidly, and sometimes expires in a few minutes. Some, however, escape in good health, and afterwards reign in the hive."—*Huber*.

When a valuable queen has been balled, prompt measures should be taken for her release. If one endeavours to break up the ball with his fingers, or with the aid of a smoker, it frequently happens that, when the outside bees disperse, one or more of those in immediate contact with the queen will sting and kill her. But if the ball be dropped into a small basin of water, it will fall to pieces; the alarm will be so great that the murderous design will be abandoned; and the queen may be rescued unhurt.

297. Use of Queen Cages.—In order to give the strange queen time to acquire the scent of the colony, and to permit the bees to become accustomed to her before her release, she should be caged on one of the centre combs containing brood and un-

Fig. 98.
PIPE-COVER CAGE.

capped honey. The pipe cover cage (Fig. 98) is the least complicated and most useful cage for the purpose. Pick up the queen, and let her run up into the cage, sliding a card underneath. Take out the centre comb: uncap a few honey cells next to capped brood: put down the cage so that it will cover some uncapped honey: withdraw the card; and press or screw the rim into the comb as far as the mid-rib, or the bases of the cells, carefully avoiding injury to the queen in any way. Do not disturb the bees again for at least twenty-four hours, or for two or three days if the colony has been long queenless. Releasing is safer if done in the evening, when the bees have quieted down. If, on releasing her, the bees on the comb show any inclination to crowd or molest the queen, cage her again until the next day. Sometimes the bees will release the queen themselves by eating through the comb; and, if a circular piece of the comb be cut from the opposite side, under the cage, and be put back again, the bees will be encouraged to release and welcome the queen in that way. Of course, there must not be another queen in the hive, and if there be any queen cells on the combs, they should be removed. The Abbott

Fig. 99.
ABBOTT QUEEN CAGE.

Queen Cage (Fig. 99) is a device by which a queen may be imprisoned in the midst of the cluster, and released without exciting the bees or uncovering the hive. The cage is slipped between two of the frames; the queen is admitted at the top; and, when the wire is drawn up, an exit at the bottom opens and allows her to pass out.

298. Introduction by Artificial Swarming.— As it is found that bees of a swarm will generally accept a new queen readily, queens are sometimes introduced by making an artificial swarm of the stock **(222),** removing the old queen, and shaking the bees off the frames before an empty skep placed on their stand. The hive is then replaced on the stand, and the bees are shaken on to a hiving board, the new queen being dropped among them as they run in. This plan is not only troublesome, but is attended with extra risks, and there is danger of having the brood chilled during the operation.

299. Direct Introduction.—It is found that a new queen can generally be introduced safely if run into the hive from above, at night, without disturbance of the stock, and when the queen is in a hungry condition, and, therefore, ready to accept food from the bees, and to show neither fear nor fight. Remove the old queen a few hours before night-fall. When darkness is setting in take out the young queen, and put her into a match-box, keeping her without food for not less than half an hour. When it is quite dark, take a lantern to the stock: quietly raise a corner of the quilt, and let the young queen run out of the match-box down among the frames: cover up: and do not open that hive again until at least 48 hours afterwards. The same match-box must not be used for another queen. This is the "direct method" introduced by Mr. Simmins many years ago, and one of the simplest and most successful methods that can be adopted.

300. Sending Queens per Post.—Queens, being sent per post, or upon any long journey, require to have some attendant bees, and a supply of suitable food *en route*. The simplest travelling box (Fig. 100), one that can be made without expense, and that

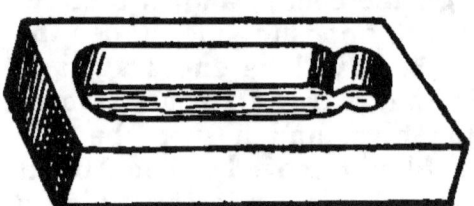

Fig. 100.
QUEEN TRAVELLING BOX.

has been used with satisfactory results, consists of a piece of soft wood 3″×1″×¾″. With a ½″ centre bit, two holes, 1″ from centre to centre, are bored nearly through the wood, and one hole with a ⅜″ centre bit. The wood between the holes is cut away, as shown; and, for ventilation, three or four holes in each side are bored with a fine bradawl. The food, consisting of honey and fine, icing sugar, as a tough dough, goes into the small hole, and the queen and her attendants occupy the remainder of the space. A piece of broken section, 3″ × 1″, makes a lid. The box is wrapped in flannel, and brown paper (ventilation being provided for), has a tie-on label, and is dropped into the post like an ordinary letter. More elaborate travelling boxes are made to serve as introducing cages also, so that the box can be introduced at once to the hive, and the queen be liberated in due course. (Fig. 100b.)

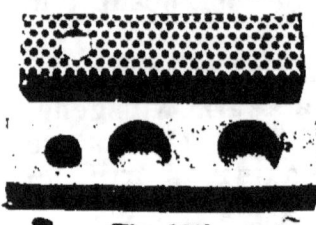

Fig. 100b.
QUEEN CAGE.

CHAPTER XXVI.

MARKETING HONEY.

301. Home Honey.—Home honey, as to its quality, can hold its own with any produced elsewhere; and we bee-keepers, close as we are to the best market in the world, should find little difficulty in disposing of our produce to advantage. But, while the English market is largely supplied by British producers, the colonies and foreign countries import at prices so moderate that the Customs Returns certify the imports of honey, into the United Kingdom, at from £30,000 to £187,000 per annum. To compete successfully in such a market, it is necessary, not only that the quality of the article be excellent, but also that the manner of presenting it for sale should place it on a level with, if not superior to, that of any other honey offered to the public. As to the quality; that may be left to the bees, and to the flowers of our unrivalled hills and valleys. As to the presentation of the article; that is a matter to which insufficient attention has hitherto been given, and which must be more carefully attended to in the future, if our honey is to attain to that position in the markets to which its quality entitles it. To the bee-keeper it is no less important than the harvesting of a large quantity of honey, that the honey should be so presented to the buyer, in the best possible condition as to quality and "make up," that the customer may desire more, and be willing to pay a fair price for it.

302. Storing Honey.—Sections, when removed from the hive, should be stored, preferably in close tin boxes, and in a dry, warm place, safe from dust, flies, mice, etc. If left in a cold, damp place, the distinctive flavour and aroma due to the essential oils of the flowers will be sacrificed, and the honey, a supersaturated liquid, will absorb moisture from the atmosphere; will become thin; will increase in bulk; and will exude through the cappings in minute drops: from which we have the too familiar "weeping section," with its whiteness and beauty gone, and its value also gone to no small extent. Or, cold may cause the honey to crystallize in the cells, which spoils it for the market, and causes many a large buyer to say, "I never purchase sections after September." Extracted honey, stored in a cold place, will granulate. It keeps best in bulk, and should be so stored until it is required for marketing **(400)**.

303. Preparing Comb Honey for Market.—Before despatching comb honey to the market, the whole stock should be gone over carefully, and graded into first, second, and third classes. This is a detail which should never be neglected, because, a few indifferent sections in an otherwise prime lot may pull down the price of the whole consignment to second, or third quality rates, thus imposing a serious loss upon the producer. First quality sections, to secure the highest price, should be of full weight, turning the scale at 16 oz.; they should be free from " pop holes," well filled and sealed, uniform, clean and attractive. Second quality sections should weigh, at least, 15 oz.; they should be joined to the wood on all four sides, and be fully sealed, except at the edges. Third quality sections will include all that are inferior to second quality, and that are suitable for sale. The wood should be scraped thoroughly clean; but, the comb being exceedingly delicate, there must be no undue pressure upon the flat surfaces of the wood, lest any of the cells attached to the section be crushed, however slightly, and a leakage occur. A slight accident of that nature may sometimes be repaired with a little piece of clean, white wax spread upon the breakage with the flat of a warm knife. If the sections are to be sold unglazed, each section should be wrapped in wax paper, the fold being made on the top of the section, the ends being turned in securely, and the package being tied with thin cord. Thus prepared, should a leakage occur in any one section, the other sections will be preserved from soiling. (See also **397-398**.)

Fig. 101.
GLAZED SECTION.

304. Glazing Sections.—Sections that are glazed (Fig. 101), and neatly finished with embossed, lace, or plain paper, present an exceedingly attractive appearance, and, being safe from flies and dust, are preferred by many retailers, who are generally willing to pay an increased price for them. The slips of paper should be exactly 17" long, and 2¾" wide. The glass should measure exactly 4¼" × 4¼", and should be cut without irregular corners. Photographers often have quantities of useless negatives, which can be purchased cheaply, and cut to the correct size. Give the paper a coat of good paste: set the section, top side up, upon it, exactly in the middle from side to side, and projecting about ¼" beyond one end of the slip.

Fig. 102.
GLAZED SECTION BOX.

Place a square of glass against each side; and turn all over on the slip, pressing the edges of the paper on the glass as each side is turned, and fastening the corners neatly with a little paste. The fold will be on the bottom; and a neat label describing the contents, and with the producer's name and address, may be pasted on the top (306). The glass should be polished clean, and the section should then be wrapped in paper. Glazed boxes for sections are much used (Fig. 102). They are inexpensive, entail little or no trouble, and are convenient packages on a merchant's counter. (See also 398, p. 213.)

305. Packing Sections for Transport.—Honey-comb sent per post or rail, requires careful packing to avoid breakage *en route*.

Fig. 103.
TRAVELLING CRATE.

The travelling crate illustrated (Fig. 103) takes one dozen sections, and has an arrangement of light springs underneath, to minimise jolting of the contents, while the glass sides disclose the fragile nature of the goods within, and appeal to the compassion of railway porters and other transport agents The Home-made Travelling Crate (Fig. 104) consists of two squares of wood, 5½" × 5½" × ½", and 12 pieces of plaster lath, 14" × 1" × ¼". The laths are tacked on three sides of the squares; two pieces of corrugated card-board, 5½" × 5½", are slipped in at each end; and the crate is well lined with straw. Six sections, carefully wrapped in wax paper, are inserted; straw is laid on top; and the remaining laths are tacked on. If this crate be corded, and a tie-on label attached, it may be sent per post with confidence. The cost of such a crate is trivial. Parcel Post Boxes (fig. 105) for sections and bottles, or jars of honey, are made of strong corrugated card-board and also of leather board. The lids are printed for addresses, and stamp tags are attached. When larger quantities are being dealt with, it is advisable to pack each dozen sections in a cardboard box 13″ × 9″ × 4½,″ to hold them close enough to prevent their moving. When filled, the box should be tied with stout cord. These boxes should travel in a strong case, 30″ × 16″ × 12″, with sides ½″ full and ends ¾″ full, 2 battens on each end outside, and 2 battens, 2″ × ½″ × 17″, on lid. The packing should be composed, not of hay, but of straw. Put a layer of two or three inches of straw on the bottom of the case: on this set some of the boxes, side by side, and with a couple of inches of straw packing between them and the case on all four sides: on top of the boxes put another layer of two inches of straw: set on more boxes: pack all round: cover with two or three inches of straw: add, on top, a note specifying the nature and quantity of the contents: screw on the lid: tie with strong rope: and, on the lid, affix a large card with the following, in distinct characters:—

Fig. 104.
HOME-MADE TRAVELLING CRATE.

Fig. 105.
PARCEL POST BOX

This Side Up.   Fragile.   Honey Comb.   With Care.

For———   From———   Date———

306. Preparing and Packing Extracted Honey for Market.—
Jars, for marketing extracted, or run honey, are made in various sizes from 2 oz. to 2 lbs. and upwards (fig. 106). The size most commonly used holds one pound. Glass is preferred generally by the public. The jars are either corked, or covered with a tie over piece of vegetable parchment, or are fitted with a metal screw cap, in which is a cork wad to prevent the escape of honey. If the honey is to be bottled, the vessel in which it has been stored should first be put standing in hot water. This causes the bubbles to rise to the surface, and helps to keep the honey from granulating afterwards. Should the honey have already become solid in the tin, the latter should be put standing for some time in hot water near the fire, until

Fig. 106.
HONEY JARS.

the contents become liquid. But the honey must not be raised beyond 160°Fahr., or its flavour will be spoiled; 144°Fahr. will be sufficient, and, as that is the point at which beeswax melts **(62)**, if a small piece of wax be put into the honey it will indicate when the vessel should be drawn back from the fire. The most secure fastening for bottles is, undoubtedly, a driven-in cork that has first been dipped in melted wax, and that, after having been driven in, is dipped in a mixture of beeswax and resin. The "tie-over," and the "screw cap" bottles are, however, generally preferred by the public. With these, a piece of waxed paper should be put on before the parchment or screw cap; and it will be an improvement to dip the cork of the screw cap in melted wax. The bottles should be perfectly clean, and an attractive label should be pasted on, as advised above **(304)**. There is a great deal more in the label, and general "get-up" of the article than many bee-men suppose. The label illustrated (Fig. 107) may be procured from Mr. E. H. Taylor; is suitable for both bottles and sections; and is printed in four colours. It is supplied for "English," "Scotch," "Welsh," "Irish," and "Heather" Honey, as required. The name and address of the producer, or of the society marketing the honey, can be printed on the label, so that the purchaser, or his friends, if pleased with the article, may have no difficulty in repeating orders. Thus, the label serves as, not only an ornament, but also a useful advertisement. As dealers frequently object to granulated honey, supposing it to be necessarily impure, it is wise to add a notice to the following effect:

Fig. 107.
HONEY LABEL.

"NOTICE.—*Honey that is pure will candy, becoming hard or crystallised, when stored in a cold place. Keep this warm. If it candies, remove the lid, and set the jar in hot water until the contents liquefy.*"

Travelling Boxes, for Honey Jars, are made to hold twelve jars in separate compartments, each compartment and the top and bottom being lined with corrugated cardboard (Fig. 108). When larger quantities are being dealt with, the bottles should be papered, and packed in a strong box, with a liberal allowance of straw. Tins for Run Honey in bulk are made to hold from 1 lb. upwards. They are fitted with lever-top lids. Those to hold 30 lbs., and over, are usually enclosed in crates which protect the tins from breakage, and are fastened at the ends by screws.

Fig. 108.
TRAVELLING BOX FOR HONEY JARS.

CHAPTER XXVII.

ROBBING AND FIGHTING.

307. Robbing.—It must be admitted that bees, notwithstanding their many excellent qualities, sometimes become very capable and persistent robbers, and that, when once this sordid vice has taken hold of them, it is exceedingly difficult to induce them to shake it off. In spring and autumn, when nectar is scarce out of doors, a careless bee-man may turn all his virtuous pets into thieving rascals, by dropping honey or syrup anywhere near the hives, or by unduly exposing it during manipulations. Then, the strong stocks destroy the weak stocks, carry off their stores, and leave them to perish of hunger. Sometimes the robbed colony, when further resistance becomes hopeless, join the robbers; and, having helped to empty their own combs, sally forth to do unto others as they, themselves, have been done by; and, the owner, paying a belated visit to his colonies, is surprised to find one or more hives empty of bees and honey **(215)**, and, upon the ground, in front, the carcases of the slain. Even when no sweets have been carelessly exposed about the apiary, weak stocks, acting as a temptation to the strong, encourage robbing, and often fall victims to it. They will, indeed, maintain a strenuous resistance against the aggressors for a time **(12)**; but this is a case in which there is safety only in numbers, and a weak colony, in such circumstances, if left without assistance, must eventually submit to defeat.

308. Precautions against Robbing.—Obviously the precautions necessary to prevent robbing are:—To avoid exposing sweets when nectar is scarce in the fields: to do all feeding in the evenings, when bees have ceased flying: and to keep all stocks strong, by uniting the weak, and by helping the well-to-do.

309. Signs of Robbing.—The signs which denote that robbing is in progress are unmistakeable. Wild excitement manifests itself about the entrance of the hive attacked: robbers hunt about the hive corners, and at all openings, seeking an entrance where there are no guards: returning bees hurry in-doors, as if in haste to escape the turmoil without; a loud buzzing is kept up without intermission: and, on the alighting board an angry fight is carried on between the robbers and the defenders of the hive; bees will be seen struggling together and rolling in couples to the ground, where one or both of them will show signs of having been injured, and where, if the fight has been fast and furious, many dead will have already fallen.

310. Treatment.—When robbing has commenced, it must be dealt with at once, or the whole apiary may be thrown into confusion. The first thing to do is to close, to one bee-space, the entrance of the hive that is being attacked; thus giving the defenders an advantage by making the enemy advance in single file. If the attack continues, one or more of the following remedies should be applied:—Place a handful of wet hay at the entrance, so that, while the bees of the hive will force their way through, the process may be too slow for robbers. Arrange on the alighting board two pieces of 1" wood, half an inch apart, and with a lath, or slate on top; so that, to come and go, bees must pass through a dark passage, which is very discouraging to robbers. Saturate a cloth with carbolic solution (Recipe 362), and spread it on the alighting board right up to the entrance, sprinkling it afresh as required. Add a little carbolic to a large pail of water, and, with a syringe or a watering pot, drench the robbers as they fly in front of the hive. Set up a piece of glass an inch from the entrance, and sloping from the alighting board to the hive front, so that the difficulty to stranger bees of finding an entrance, may be increased. If none of the remedies described proves effectual, close the hive entirely until the evening, opening the doors to full width, covering the entrance with perforated zinc, and taking care to give all the ventilation required (218). It may even become necessary to remove the molested hive from the apiary altogether until the danger is over. When the robbing is being carried on by the bees of only one hive, some apiarists deal with it by flouring the bees as they leave the attacked hive, thus discovering the colony from which the robbers come; after which they transfer the hive of the robbers to the stand of the robbed, and *vice versa*, until the mischief ceases.

"In Germany, when colonies in common hives are being robbed, they are often removed to a distant location, or put in a dark cellar. A hive, similar in appearance, is placed on their stand, and leaves of wormwood and the expressed juice of the plant are put on the bottom board. Bees have such an antipathy to the odour of this plant, that the robbers speedily forsake the place, and the assailed colony may then be brought back. The Rev. Mr. Klein says, that robbers may be repelled by imparting to the hive some intensely powerful and unaccustomed odour. He effects this the most readily by placing in it, in the evening, a small portion of musk, and on the following morning the bees, if they have a healthy queen, will boldly meet their assailants. These are non-plussed by the unwonted odour, and, if any of them enter the hive and carry off some of the coveted booty, on their return home, having a strange smell, they will be killed by their own household. The robbing is thus soon brought to a close."—S. WAGNER.

It will often be found that a colony which offers little resistance to robbers, and is overpowered, is either queenless or diseased; and that bees that are being robbed are more than usually difficult to handle. (179). If it should become necessary to remove the molested hive, it will be well to place an empty hive, of similar appearance, in its place, as otherwise the robbers may attack the neighbouring stocks. Sometimes a useful remedy consists in inserting a "Porter" escape (274), or a double cone (273) in the entrance of the hive that is being attacked-any other inlet being closed—and to leave it so for forty-eight hours, thus confining the bees to the hive, but giving sufficient ventilation over the frames by placing on them a sheet of wire cloth, and removing the other coverings. After forty-eight hours, the entrance being opened, robbers and robbed are frequently good friends and work together in harmony; and the stock, weak before, but now reinforced by the robbers, will probably be strong enough to defend itself.

CHAPTER XXVIII.

FEEDING BEES: RECIPES.

311. Objects of Feeding Bees.—It is a common notion, and a very mistaken one, that bees, being so well able to forage for themselves, require no artificial feeding. During several months of the year there is little or no nectar to be gathered in the fields: sometimes when natural food might be had in abundance, stress of weather confines the bees to their hives, so that they cannot visit the flowers: in winter and early spring, foraging is impossible: and, frequently, when the bee-keeper has taken his harvest from the hives, the bees are left without sufficient food to carry them through the cold months. In such circumstances, neglect to supply food artificially is often accountable for the death of many stocks. And, by feeding bees, there are other objects to be gained beyond that of staving off starvation. The general desire to obtain a large harvest of honey can be satisfied only by having the stocks as strong as possible before the honey flow opens **(255);** for, only the bees that have been born at least fourteen days, from eggs laid at least thirty-six days before the honey flow opens, can take full advantage of it. **(190).** But, it is found that neither will the queen put forth her best laying powers, nor the bees consent to rear brood in quantities, until food begins to come in abundantly **(192).** Similarly, towards the close of autumn, it is necessary to have a large quantity of bees reared to survive the winter, and to carry on the work of the colony in the spring. But, with the cessation of the honey flow, breeding will naturally decrease, unless food be supplied; and, even the eggs and larvæ will be destroyed when food becomes scarce, with the result that the stock may come out in the spring too weak to be of any practical use in the season following. Therefore, if good results are to be secured, nature's supply must be anticipated, and supplemented, by artificial feeding: wise and timely attention to this detail may make all the difference between a good and a bad, or indifferent honey harvest **(202).** It goes without saying that sugar is not as good feeding for bees as is honey. Experienced bee-keepers are careful not to deprive their stocks of more honey than the bees can afford to give, having regard to the needs of the colonies.

312. Precautions.—The following precautions, as applicable to the feeding of bees, should be adopted as rules for invariable observance:—(1) Use only pure, refined, cane sugar: other sugars are injurious to bees **(330).** (2) Never permit the sugar to become burned during cooking: even pure, refined, cane sugar, if burned, will do much harm, especially in cold weather when bees are confined to their hives. (3) Contract the entrances of all hives in which feeders are being used, and do not allow robber bees access to the food: robbing is often set up through neglect of this precaution **(307).** (4) Give the food warm, in the evenings, when the bees have ceased flying: bees will frequently refuse cold syrup in spring and autumn. (5) Keep all feeders warmly covered. (6) Never leave supers on a hive when sugar-feeding is in progress in that hive: syrup stored in sections or extracting supers, will render the honey therein unfit for sale. (7) Do not hesitate to spend money on sugar: it is only quarter the price of the honey you get instead.

313. Spring Feeding.—Bees are fed in spring, and at other seasons, to "stimulate" them **(192)**, and at all times when it is found that their stores are insufficient. Except in winter, when candy is the food employed, syrup is given. Spring feeding begins when the bees begin to fly freely—in March or April, according to the season and locality. Honey in the combs may have a couple of inches of cappings bruised once a week, exposing the food for use. In early spring, when the nights are cold, bees will often refuse to take down syrup. In such a case, if there be any liquid honey at hand, a good cake of candy may be made by mixing honey with loaf sugar pounded fine, and the cake may be put on the frames, under the sheet, so that the bees may easily reach it. Flour candy (Recipe **324**) forms an excellent food for bees in spring, and stimulates brood rearing to a surprising extent. Liquid food may be prepared according to the directions given at the close of this chapter (Recipe **321**). The supply should be regulated according to the season, the needs of the colony, and the objects in view. In spring, for stimulative purposes, i.e., to induce more rapid brood rearing, the supply should be very gradual—say a wineglassful given through two or three holes only, and that, during the night, the supply being cut off in the morning; for, a supply too rapid will lead to the storing of syrup in the combs required for brood, and this is to be carefully avoided in the spring. With this object, feeders are employed which introduce the syrup immediately over the cluster, and which permit the supply to be regulated according to the requirements **(120-123).**

Artificial pollen should also be given where a natural supply is wanting. (192).

314. Summer Feeding.—Feeding in summer becomes necessary during a spell of bad weather, and is often desirable between the early honey flow from fruit trees and the main flow from clover, and also between the latter and the heather flow. Swarms should be fed for a few nights to the extent of half a pint of honey or syrup per night, to assist them in drawing out foundation into comb, and to prevent the danger of hunger, resulting in the cessation of breeding, the throwing out of immature brood, and the dwindling of the swarm. But, swarms to which have been transferred the supers from the parent stocks (240) do not generally require feeding, and should not be fed, while the supers are on, except with honey (312). When the honey flow ceases and supers have been removed (272), liquid food may be given again, and this may be continued, to the extent of about half a tumblerful per night, until the middle of September. As a result, breeding will be continued uninterruptedly, and a large supply of young bees will be reared to maintain the colony in the winter, and to begin work in the spring (202). Syrup for summer feeding is made similar to that used in spring (Recipe 321). In very bad seasons, when the bees cannot procure natural pollen, flour candy will greatly assist breeding in the hive (Recipe 324).

315. Autumn Feeding.—Autumn feeding begins in the middle of September, and is intended to supply sufficient food to carry the stocks through the winter and early spring, if there is not an adequate quantity of honey in their combs. It should be given warm, every evening, and as rapidly as the bees can be induced to take it down, so that it may be stored and sealed in the combs before cold weather arriving renders the capping of the cells impossible (377). Unsealed stores are liable to ferment, and such food is highly injurious to bees (330). The syrup should be thicker than that used earlier in the year, and may be made according to the directions given later on (Recipe 322). A colony, to winter safely, should have, at least, 30 lbs. of sealed stores. Six standard frames (97), well filled, will suffice, and no strong colony should be considered safe with less. A Dutchman, when asked—"How much beer is enough for a man?" is said to have replied—"Too mush peer is shust enough." More accurately it may be said that too much food is just enough for bees in winter. As the object of autumn feeding is, not to encourage breeding, but to rapidly supply stores for winter, the feeders used in spring and summer are not invariably suitable in autumn. (123-125). In an emergency, when there is not time for supplying autumn syrup through a feeder, empty combs may be carried into the house, and the warm syrup may be poured direct into the cells, until both sides of the combs are filled. The combs may then be carried out in a comb box **(173)**, and inserted in the hives requiring them. But this method, also, must be adopted, if at all, sufficiently early to admit of the capping of the cells before the arrival of cold weather. It is an excellent plan to use one or more stocks to store and seal the syrup for all the other stocks, as previously advised. **(124)**.

316. Winter Feeding.—When stocks are short of food in the winter, only sealed honey, or candy can be given with safety.

"Experience shows that stocks, no matter how well supplied with food below, winter better when they have a cake of candy on top of the frames. The bees use the candy first; and, when they have consumed a little of it, they have a safe winter passage across the frames. Every bee-keeper who is not quite certain that his stocks are sufficiently supplied, should give them 'the benefit of the doubt,' in the shape of a cake of candy—candy not hard enough to require a pickaxe to break it, but candy that is properly made, soft, and palatable, and good. (It may be made according to Recipe **323**). Let it cool for half-an-hour. Then, gently slip a cake under the sheet of each hive, so that the candy shall be directly over the clustering bees. Renew the supply of candy as required. Pressure of the fingers on the sheet will show when the candy has been used. A neater plan for supplying the candy, and one that will repay the little extra trouble, where only a few hives have to be dealt with, may be adopted as follows. Procure for each hive a small, shallow box of wood, or cardboard; remove the lid and cut, in the bottom, a hole to correspond with the hole in the sheet that is on the frames. Put a piece of newspaper over the hole in the bottom of the box, and fill up with candy. Now, set an empty section crate on the sheet that covers the frames; pull the paper off the candy; and set the box on the sheet, so that the bees shall have access to the candy right over the cluster. Place a piece of glass on the box. Fill up the crate with warm stuff, such as tailors' cuttings, cork dust, or chaff; pack all round it with cloth or newspaper; and set the usual quilts on top. Thus, there will be no escape of heat; the candy will be in the warmest part of the hive; and the glass will enable you to see when a further supply of food becomes necessary."—J. G. D. in the *Irish Bee Journal.*

317. Feeding for Comb Building.—It has already been pointed out that careful bee-keepers make it a rule to have empty combs always at hand when required **(193)**. There are certain weeks, in every year, when bees are comparatively idle, during a cessation of nectar-secretion in the flowers: the opportunity may then be taken advantage of to procure new combs for future use. If frames of foundation be inserted alternately with the brood combs, and if thin syrup (Recipe

321) be given, the bees will fill the frames with comb in an incredibly short time. Care must be taken to withdraw the new combs before the queen shall have begun to oviposit there: or, failing this, they may be left in their position for brood rearing, and a corresponding number of broodless combs, if any are there, may be withdrawn and packed up in a dry place until required. Such combs will be of immense service if given to swarms, and will, in other cases also, effect a saving of valuable time in the height of the season.

318. Feeding Bees in Skeps.—Stocks in skeps may be considered safe for winter if the skep, on being weighed, is found to exceed 25 lbs. When syrup feeding in skeps is necessary, it should be given overhead. An ordinary skep may have a hole cut in the top sufficiently large to admit the mouth of an Economic, or a Bottle Feeder **(120-121).** When the feeder is placed in position, it should be wrapped round with warm material, to prevent the escape of heat, and a cover, such as an empty skep, or a large flower pot with the hole stopped, should be put on, to shut out prowling, stranger bees. In autumn or winter, a bar of candy may be pushed into the hole, and covered up. A better plan is to cut two or three inches off the top of the skep, and to put on, instead, a piece of board with a hole in the centre, over which a bottle and stage feeder **(121-122),** or a cake of candy may be placed. The board should be fastened securely by nails passing through it into the skep, and a safe cover should be put on over the feeder. Flat-topped skeps **(271),** made to take supers, can have feeders placed on them in the same manner as described above for modern hives.

319. Water.—Bees cannot carry on their wonderful work without water **(9).** If they have not access to natural sources close enough to their hives, water should be supplied to them. It is neither necessary nor desirable to add salt. A vessel of water, with corks floating in it on which the bees may alight: a tumbler of water inverted on a plate: or, a bowl of water with a sponge, or a piece of cotton wool in it, through which the bees may suck up what they require, will serve the purpose. The vessel should be placed in a sheltered, sunny spot. Bees will often resort to dirty pools of stagnant water, rather than take clean, but colder water from an artificial source. (Fig. 113.)

320. Pollen.—The necessity for supplying artificial pollen when a natural supply is not available, has been referred to elsewhere, and directions for supplying it have been given under the heading—" Stimulating in Spring." **(192).**

RECIPES FOR FEEDING.

Pure, refined, cane sugar only, to be used as follows:—

321. Spring and Summer Syrup (160, 313, 314, 317).—

1 part hot water, by weight, to 1 part sugar, by weight, thus:—

Water ...	1 pint	1 quart	2 quarts	4 quarts	12 quarts
Sugar ...	1¼ lbs.	2½ lbs.	5 lbs.	10 lbs.	30 lbs.

Stir incessantly over a slow fire until the sugar is dissolved. If foul brood be feared, add Naphthol Beta Solution (**325**), or Izal, or Yadil, to the above quantities, while the syrup is hot, as follows:—

N. Beta ...	½ Teaspoonful	1 Teaspoonful	1 Dessertspoonful or ¼ fluid oz.	1 Tablespoonful or ½ fluid oz.	3 Tablespoonfuls or 1½ fluid oz.
Yadil ..	2½ ditto	5 ditto	5 ditto	5 ditto	7½ fluid oz.
Izal ..	8 drops	15 drops	½ Teaspoonful	1½ Teaspoonfuls	1 Tablespoonful or ½ fluid oz.

(For measures—see par. **326***.)*

322. Autumn Syrup (315).—

1 part hot water, by weight, to 2 parts sugar, by weight, thus:—

Water ...	1 pint	1 quart	2 quarts	4 quarts	12 quarts
Sugar ...	2½ lbs.	5 lbs.	10 lbs.	20 lbs.	60 lbs.
Vinegar ...	1 Dessertspoonful or ¼ fluid oz.	1 Tablespoonful or ½ fluid oz.	2 Tablespoonfuls or 1 fluid oz.	1 Wineglassful or 2 fluid oz.	3 Wineglassfuls or 6 fluid oz.

Stir incessantly over a slow fire until the sugar is dissolved, and allow the syrup to boil. If foul brood be feared, add Naphthol Beta Solution (**325**), or Izal, or Yadil, to the above quantities, while the syrup is hot, as follows:—

N. Beta ..	1 Teaspoonful	1 Dessertspoonful or ¼ fluid oz.	1 Tablespoonful or ½ fluid oz.	2 Tablespoonfuls or 1 fluid oz	1½ Wineglassfuls or 3 fluid oz.
Yadil ..	5 ditto	2½ Tablespoonfuls	1¼ Wineglassfuls	2½ Wineglassfuls	15 fluid oz.
Izal ..	15 drops	½ Teaspoonful	1½ Teaspoonfuls	3 Teaspoonfuls	2 Tablespoonfuls or 1 fluid oz.

(The Vinegar is added to prevent, or retard, crystallization.)

(For measures—see par.* 326.*)

323. Candy for Winter Food (316).—

1 part hot water, by weight, to 5 parts sugar, by weight, thus :—

Water ...	½ pint	1 pint	1 quart	2 quarts	5 quarts
Sugar ...	3 lbs.	6 lbs.	12 lbs.	24 lbs.	60 lbs.
Cream of Tartar	½ Teaspoonful	1 Teaspoonful	2 Teaspoonfuls or ¼ oz.	1 Tablespoonful or ½ oz.	2½ Tablespoonfuls or 1¼ oz.

Boil the water, withdraw it from the fire, add the requisite quantities of sugar and acid, as above, stirring until dissolved. Return it to the fire and *stir it unceasingly* until it boils, for if you allow the sugar to burn, through lack of stirring, it will be as poison to the bees **(312)**, and your ingredients, time, and perhaps patience, will be lost. Continue boiling and stirring until the mixture begins to thicken; then test it by dropping a little on a cold plate. It must set soft, but not sticky; if it sticks to your fingers, boil it a little longer. When it proves right to the touch, cease boiling it, and if foul brood be feared, add Naphthol Beta Solution in quantity as directed below. Without delay, stand the vessel in another containing cold water, and stir vigorously until the mixture begins to set, when you must *immediately* pour it into receptacles previously prepared for the purpose; for the mixture will set quickly, and must be dealt with promptly. The receptacles may be saucers, or soup plates, on each of which a sheet of strong paper must be laid to receive the candy; or shallow boxes may be used **(316)**. If paper be adopted, it will be well to lay upon each cake two pieces of ¾" stick, 1" apart, and parallel, pressing them into the candy before it cools, and flush with its upper surface, so that when the candy shall have been consumed, the " winter passage " **(377)** may be preserved by the sticks *lying across the frames*. If preferred, frames may be prepared with four or five lengths of stout string, after the manner of wires in frames **(262)**, but running vertically, and into these frames, lying on waxed or slightly greased paper, the mixture may be poured. When the candy is cool, a cake may be given under the sheet and quilts, or a frame of candy—the paper having been removed—may be inserted next the brood nest, in any hive requiring it. To medicate with Naphthol Beta **(325)**, Yadil, or Izal, stir in the quantity required while the mixture is still at boiling point, allowing to 3 lbs. sugar 1½ teaspoonfuls N. B. Solution, 1 teaspoonful Yadil, or 20 drops Izal, and to other weights of sugar in like proportions.

CHAPTER XXIX.

DISEASES, &c.

327. Diseases, &c.—Bees are subject to various diseases and ailments, among which may be included—Dysentery; Paralysis; Chilled Brood; Black Brood; Pickled Brood; Foul Brood, or Bee Pest, and a new disorder, commonly called "Isle of Wight Disease" **(360)**. In the treatment of these, modern bee-keeping enjoys a distinct advantage as against the old-fashioned methods **(77)**; for, whereas the skeppist, being unable to make thorough examinations of his stocks, could but seldom discover an unhealthy condition before the disease had made considerable progress, the moveable-comb hive enables the bee-keeper to discern the first approaches of danger **(81)**, and, by the use of preventives and remedies, to restrain disease, or to cure it in its initial stages. With this object in view, it is important that, when stocks are being manipulated, a sharp look-out be kept for any signs of disease; that preventives and remedies be always at hand when required; and that, when sickness of any kind shows itself, immediate steps be taken to deal with it.

328. Dysentery.—When bees are suffering from dysentery, the ailment will show itself at the close of winter, or early in spring.

329. Symptoms.—On examining the stock affected, it will be seen that the bees have discharged their excrements over the combs, and on the sides, floor, and alighting board of the hive, as they never do in a healthy state, being scrupulously clean in all their habits. **(11)**. The fæces have a very offensive smell, and vary in colour from a red-brown to a mud-black, according to the nature of the food that has been used. The bees move about languidly, and the colony rapidly dwindles.

330. Cause.—When bees have been long confined to their hives, and unable to take a cleansing flight: when they have, from any cause **(378)**, such as untimely manipulations, consumed an excessive quantity of food: or, when their food has consisted, to any considerable extent, of sour, or unripened honey **(315)**, or of syrup made with unsuitable sugar **(312)**, they become subject to dysentery; and, being unable to retain the excrements, they void them anywhere **(377)**.

331. Prevention.—To guard against this complaint, late manipulations, causing undue excitement and consumption of food, and late feeding with syrup, when evaporation and sealing of the food are impossible, should be avoided: none but pure, refined cane sugar should be used for syrup and candy feeding.

332. Treatment.—When an attack of dysentery has set in, the bees should be transferred to a clean hive, contracted to the space occupied by the cluster; very soiled combs should be removed and washed clean, and their places should be occupied by clean combs; candy, or sealed honey should be given; and, the bees should be kept warm, and as free as possible from excitement. A few warm days generally put matters right.

333. Paralysis.—Bee paralysis is a disease which exists to some extent in this country, although it is not often reported. A peculiarity of the disease is, that it comes and goes in an unaccountable manner, suddenly attacking a strong colony, and reducing it to the condition of a nucleus, and sometimes disappearing as suddenly, leaving no apparent trace behind, save the depleted state of the stock.

334. Symptoms.—In the early stages, the affected bees will be seen leaving the hive, their abdomens greatly swollen. Later on, the trembling, or shaking paralysis shows itself. The healthy bees seize the unhealthy, and drag them from the hive; no resistance is offered, and, in an incredibly short time, the stock will become small and weak, and will, if the disease continue, be wiped out. (360).

335. Treatment.—Among the remedies prescribed are (1) requeening; but, in pronounced cases, a change of queens has little effect: (2) transferring the stock to the stand of a strong, healthy stock, and *vice versa*; thus providing a force of strong, healthy bees to remove the diseased and infectious bees to a distance: and, (3) dusting with sulphur, which is said to show good results when thoroughly done. This method is, to remove all combs containing brood or eggs, giving them to another colony; and, in the same evening, when all the bees are at home, to dust every comb and every bee in the hive with sulphur. On the next day, the combs previously removed are returned to the hive; the reason for their removal in the first instance being, that dusting them with sulphur would kill all the unsealed larvæ, and would also kill all larvæ hatched in them subsequently. If the combs that have been treated be given to strong colonies, the bees will clean out the cells, and no mischief will result. No evidence of a cure will show itself before a week or ten days have elapsed; therefore, the treatment is not to be considered a failure when good results are not immediately visible.

336. Chilled Brood.—"Chilled Brood" is the name given to the condition of larvæ which have died through lack of the heat necessary to their life and development. It is frequently found after any sudden decrease of temperature out of doors, and in the apiaries of careless, or ill-informed bee-keepers.

337. Symptoms.—Chilled brood is sometimes mistaken for foul brood **(349)**, but, examination of the contents of affected cells will show the larvæ, in the case of chilled brood, grey in the initial stages, and subsequently black, whereas brown is the colour assumed by foul brood **(350)**.

338. Cause.—Chilled brood may be caused in spring, when the brood nest has been extended, by a sudden return of cold weather forcing the bees to cluster in the centre of the brood nest, and to leave the outer patches of brood uncovered: it may result from premature or excessive "spreading of the brood" **(193)**; or from undue exposure of the brood combs during manipulations. **(185b)**.

339. Prevention.—To avoid the danger of chilled brood, hives should be kept warm during spring and late autumn breeding: spreading the brood should be practised with much discretion, full account being taken of the prevailing weather and of the risk of a sudden drop in the temperature at night: manipulations of the brood nest on cold days should be avoided as much as possible: and, at no time should combs containing brood be exposed to chill winds.

340. Treatment.—When time permits, the chilled brood may be picked out of the cells with a pin, and be buried. Failing this, and where a large number of stocks have to be dealt with, the cappings of cells containing chilled brood may be broken, when, the bees will carry out the dead.

341. Black Brood.—In America, this disease is now identified with "European Foul Brood," which is the name there given to the disease investigated by Cheshire **(351)** and attributed by him to *Bacillus alvei*. It is now ascribed to the germ *Bacillus pluton*. It is highly infectious in larvæ, but not in adult bees **(359b)**.

342. Symptoms.—The brood is usually attacked in the early larval stages, and death generally occurs before the cells have been sealed. A yellow, pin-head spot on the larva, is the first sign, and death resulting, the larva becomes brown in colour, and finally almost black. But, whereas the rotten masses in foul brood become sticky, and ropy, in black brood they turn into a granular, liquid condition, not adhering to the cell-walls, and having a sour smell quite unlike that given off by foul brood.

343. Cause.—B. pluton: Infected food, or infected combs conveying the disease to the colony, and to other colonies, through the agency of robber bees.

344. Treatment.—Unite weak stocks, in clean hives, and on starters of foundation, confining the queen in a cage that will admit of her being fed: a few days later, substitute full sheets of foundation for the starters, keeping the queen caged for a few days longer: feed from the outset, with medicated syrup (Recipes 321, 322): render all infected combs and starters into wax by the boiling method (280): and disinfect all hives and appliances which have been in contact with the disease. (Recipe 363).

345. Pickled Brood.—Pickled brood is not prevalent in this country, nor is it nearly so contagious, infectious or destructive as either black brood or foul brood.

346. Symptoms.—Like black brood, this disease is frequently mistaken for foul brood; but, the symptoms are too distinct to admit of any doubt upon the part of a careful observer. The dead larva, generally much swollen, lies on its back, with both ends upwards: it is first white, like healthy brood, afterwards changing to yellow, gradually darkening until it is nearly black: it is never sticky or ropy: and the larva being "pickled" in its own liquids, putrefaction is arrested, and the evil smells so characteristic of advanced black brood and foul brood, are entirely absent.

347. Cause.—The disease, which is infectious and liable to be carried from hive to hive by robbers, is due to a white fungus growth starting a ferment in the alimentary canal of the larva.

348. Treatment.—The treatment commonly adopted is, to transfer the bees to clean hives, with frames of foundation; confining them to their hives for three days, so that all the infected material may be consumed; and feeding them with medicated syrup. (Recipes 321, 322).

349. Foul Brood.—Foul brood (*Bacillus alvei*) is a specific infectious disease caused by bacteria, and, next to "I. W.", the most serious disease to which bees in this country are subject. It attacks adult bees as well as larvæ, and is so exceedingly virulent that, if not speedily brought under control, it destroys colonies, devastates whole apiaries, and reaching out to unaffected places, spreads death and destruction far and near. When once it has taken possession of a district, the difficulty of thoroughly eradicating it is so great, and its contagion is so active, that entire parishes and counties may become affected to such an extent as to render bee-keeping therein an impossibility.

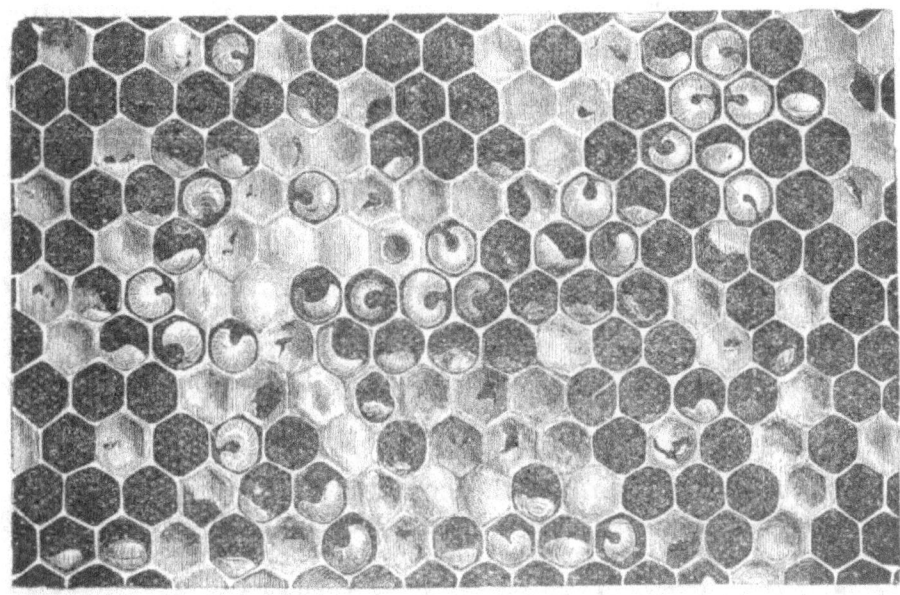

a   a
Fig. 109.
COMB INFECTED BY FOUL BROOD.

350. Symptoms.—The first signs of the presence of foul brood are manifested in the larvæ from the age of one to five days. At that age, healthy larvæ occupy the combs in regular patches of brood, the larvæ being pearly white, and lying on their sides, curled up in shape of a crescent, at the bottoms of the cells. When disease sets in, the larva assumes a different position (Fig. 109a); loses its plumpness and whiteness; and takes an unhealthy buff, or yellow tint which, as the disease developes into death, changes to brown. Then follows decomposition; the mass settles in the bottom of the cell as a rotten, glutinous, coffee-coloured matter which frequently gives off an intolerable stench. This stench is not invariably present; but it is seldom entirely absent, and, in cases of full development, it may often be detected some yards from the affected hives; it resembles the smell given off by old, melted glue; and, once experienced, is never forgotten. The bees do not appear to make any attempt to carry out the foul matter when once it has reached the point of decomposition; sometimes they fill up the cell with honey, covering the foul matter beneath, and thus contaminating the food which, when supplied to larvæ, infects and kills. Finally, the putrid mass shrinks, and clings to the lower side of the cell in the form of a stiff, black scale. Larvæ, attacked at a later period of their growth, and sealed up in their cells, die, decompose, and turn to dry scale in the same manner. These cells will remain closed when adjacent cells, having given birth to healthy brood, are open; and this, in itself, will often be sufficient to arouse the suspicion of the owner. The cappings of such cells will be seen to be darker than those covering healthy brood, and to be, in some cases, indented, as if pressed with a pencil point, and sometimes, even perforated with jagged holes (Fig. 109). If one of those cells be opened, before the contents have reached the scale stage, and, if a pointed stick be inserted and withdrawn slowly, it will bring out the sticky, elastic, brown mass which is an unmistakable indication of the existence of foul brood. Adult bees, suffering from the disease, die off very rapidly; and the remainder lose heart, become listless, and loiter about their unhappy and unhealthy home; or, fanning at the entrance, try, in vain, to remove the fetid air which they seem to recognise is, for them, the precursor of doom. When any of the symptoms described are noticed, an immediate examination of the combs should be made; and, if dead brood be found, the other symptoms should be looked for, with a view to discovering whether the mischief present is due to foul brood, or to either of the other diseases described above.

351. CAUSE.—"*Bacillus alvei*," is the name given by Cheshire, in 1883, to the rod-shaped, pathogenic micro-organism causing foul brood. (359b). Dr. G. F. White (U.S.A.) attributes it to *Bacillus larvae*.

"The *bacillus alvei*, which interests the bee-keeper, is of medium size, rod-like in shape, and four times longer than it is broad; and it would take one hundred and twenty-eight billions of them to equal a worker bee in size. If we placed a bacillus and a bee along side of each other, and wanted to place a body along side of the bee as much larger than the bee, as the bee is larger than the bacillus, we should have to place a house two hundred feet long, one hundred feet wide, and over fifty-seven and a half feet high; and, if we wished to go on and keep up the proportion, we should require one hundred and twenty-eight billion houses for the next body. They grow and multiply with wonderful rapidity. They divide by budding, or transversely across their length every hour, and if one bacterium could keep up this division for three days, it would convert over seven thousand tons of organic matter into bacteria. They form, under certain conditions, spores, or seed-like bodies which can withstand boiling water for one or two hours."—A. W. SMYTH M.D, in the *Irish Bee Journal*.

In the early stages, bacilli only are present; but later, spores are produced in enormous numbers—billions of them in one dead larva, and more exceedingly minute than the dust particles visible in a sunbeam when it shines through a chink in a closed shutter. These minute spores may be carried in the air, on the bodies of robber bees, or on the person of the bee-keeper, from hive to hive, or from apiary to apiary, to infect other stocks, and to set up disease in hitherto healthy localities. They get into the honey, and are fed with it to the larvæ; then follows a brief period of incubation, after which the bacilli are produced, which feed upon and destroy the larvæ, and pass into the spore state, to re-appear in the resulting bacilli. The spores are more dangerous than the bacteria, because of their wonderful powers of resistance to treatment which would speedily overcome the bacilli. Cheshire declares that he found the bacilli in queens, and, not only in their organs, but also in the partially developed eggs in their ovaries. They are capable of growing in any favourable medium; but bee-larvæ, as it is for them a richer soil, offers special attractions. Weak colonies, and such as are living upon unhealthy food, or in insanitary conditions, are generally the first to be attacked, so that, often, at the outbreak, it is not, as is sometimes supposed, the disease that has weakened the stock, but the weakness of the stock that has invited the disease.

352. PREVENTION.—Foul brood is eminently a disease to which may be applied, with special force, the maxim—"Prevention is better than cure." For, while the cure must always be exacting and anxious, and to some extent uncertain, the disease may generally be prevented by methods which, while they involve little trouble to the bee-keeper, are, in many respects, of incalculable benefit to his stocks. In the forefront of all desirable precautions may be placed—cleanliness; the elimination of all weak stocks by uniting; and the encouragement of strong colonies by the use of only young, vigorous queens, and suitable food (321-5); thus opposing to the assault of the bacteria, the vigour of stocks qualified, by a healthy constitution, to resist the disease.—

"The bees, when their colony is favourably situated, can resist the disease to a great extent, and the stronger the colony the greater is the resistance. In the treatment of infectious diseases in man and animals; and in experiments made by inoculating animals with parasitic bacteria, the only way yet found to save the infected animal is by strengthening and increasing the resistance of the host, so that the parasite and its poison may be unable to prevail against it. The best and safest germicides in foul brood are the bees themselves. If we cultivate the bees more and the bacteria less, spores will not be so abundant in the hive, and the bees will be able to attend to them."—A. W. SMYTH, M.D., in the *Irish Bee Journal*.

Naphthaline—an intestinal antiseptic and parasiticide, acting as a disinfectant to arrest decomposition, enables, or encourages the bees to remove diseased larvæ from the hive before decomposition sets in, at which point they refuse to do so. Naphthaline is supplied in balls; and, two of these, divided into four parts, are placed on the floor board, in the corners, of the brood nest. These pieces disappear by evaporation in a couple of months, and are sometimes covered over with propolis by the bees in order to suppress the smell, to which they have an objection. This preventive should be renewed as required. Naphthol Beta, or Izal, should be added to syrup and candy used for feeding; they are powerful disinfectants and intestinal antiseptics, very effective in parasitic diseases; N. B. has a sharp, pungent taste, and an odour resembling Phenol; it is supplied in powder, and is soluble in alcohol (325). Hives, and all appliances used in the apiary, should be kept in a condition as unfavourable as possible to the bacteria: cleanliness, a lesson taught by the bees themselves, should be a fixed rule of management: the moveable floor board should be a *sine qua non* in every hive, and should be cleaned and disinfected frequently: and, the bees should have liberty to carry on their sanitary work in every part of the hive in which organic matter favourable to the growth of bacteria may be located.

353. Treatment.—Foul brood is frequently looked upon as an incurable disease, to be ended only by sulphur and fire. But, it has been established beyond doubt that the disease can be cured, if taken in its initial stages, and even when the attack has developed considerably, if patience, perseverance, and thoroughness, with sufficient knowledge of the proper methods to be adopted, be brought to bear upon it. Destruction by fire need be recommended only when the disease has been allowed to make such headway that the stocks affected have been reduced to a condition that renders them not worth saving; or, when the bee-keeper has no qualified friend to help him, and is, himself, either too inexperienced, too indifferent, or too lazy to undertake a systematic and, perhaps, protracted cure. In such a case, it will be better to burn the lot out of hand, than to suffer weakened colonies of diseased bees, and hives that are infected, to attract robber bees from healthy colonies, and to scatter infection throughout the district.

"Rational and simple cures for foul brood have been so long known to many practical bee-keepers, that it seems strange there are others quite unable to cope with the disease when it makes its appearance in their apiaries. The disease has been cured in the past, and can as readily be cured to-day. There is really no excuse whatever for the continued existence of foul brood in any apiary, in the light of facts already placed before bee-keepers."—*Simmins.*

354. Early Stages: Treatment with Formalin.—Formic Aldehyde, produced by the limited oxidation of Methyl

Alcohol: a gas, condensible by cold to a clear, mobile fluid. Formalin, the commercial article, is stated to be a 40 per cent. solution; a powerful antiseptic, and caustic: the vapour is irritating to the eyes and nose: the article should be used with caution. When the disease is discovered in its early stages, that is before it has reached the spore stage, it may be treated with formalin as follows:—Make a solution of one part formalin to four parts of water (Recipe 365). Procure a syringe, or a glass and rubber "filler," such as is used with fountain pens, and a piece of pointed stick. Remove a frame of affected comb, and shake the adhering bees back into their hive: break, with the stick, the cappings of the diseased cells, and, with the filler, inject a drop or two of the solution into each of such cells. When the stick is not actually in use, keep it in the bottle of solution, and, at the close of the operation, burn it, and wash the filler with solution before putting it aside for future use. Next take a little of the solution, add twice as much water (Recipe 366), and with this new solution saturate a piece of cloth, or wool, and place it on the floor board: or, if you have a ventilator in the floor board (85), place the cloth underneath the perforated zinc, so that the fumes may ascend into the hive, (Illus. p. 197), and renew the application, from time to time, as required. Disinfect the clothes and hands immediately afterwards, lest you should carry infection to other hives (Recipe 364). If there are no supers on the hive treated, feed gently with medicated syrup (Recipe 321), which will be used by the nurse bees in feeding the larvæ. This remedy is a simple one in its application, and has been proved to be most useful, when adopted in the early stages of the disease.

355. Advanced Stages: Treatment by Burning.—It frequently happens that bee-keepers, who are not familiar with the early symptoms of the disease, do not discover its presence until it has too far advanced to be successfully treated as recommended above; that is, until there are present, not only single, scattered, infected cells, but also uneven quantities of diseased brood, with cells indented or perforated, and containing the coffee-coloured, ropy mass described (350). In this and subsequent stages, remedies, to be effective, must be thoroughly and continuously applied; and, as has been said above (353), if the stocks have been reduced to a condition of uselessness, and if the owner is not prepared to tackle the disease in a patient, determined manner, it will be wisest for him, and more humane, to smother the bees, and to burn all the contents of the hive; thus, by killing the spores and destroying the infected combs, etc., protecting his healthy colonies. This should be done in the evening, when there are no bees flying. The smothering of the bees may be best accomplished by the use of sulphur, or bi-sulphide of carbon. Remove couple of sods beside the hive, and open a hole about a foot square and, say, three inches deep. In the centre of this hole place the lid of a small tin box, and into this lid put about a tablespoonful of sulphur. Drop a red coal into the sulphur, and immediately lift the hive from its floor board and set it down over the sulphur, taking care that no bees escape between the hive and the ground. In a couple of minutes, the bees will be dead. Now, thoroughly saturate the mass with petroleum, and set it on fire; and when all has been burned, throw in the earth, and put the sods in their places. To kill with bi-sulphide of carbon:—Close the doors of the hive: separate two frames, and push down between them a piece of tow, cotton, or wool: on this pour a tablespoonful of bi-sulphide of carbon: drop a lighted match upon it, and immediately set on the quilts and roof. In less than a minute, the bees will be dead. There will be a slight explosion when the lighted match comes in contact with the bi-sulphide; but there will be no danger to the operator, if he be careful to keep his head back from the hive. It must be said, however, that bi-sulphide of carbon is a highly inflammable substance, and should be handled always with extreme caution. When the bees have been smothered, they can be brushed into a hole, burned, and covered with the earth and sods. The frames, combs, sheet, quilts, and all hive fittings that have been in contact with the disease, should also be burned and buried. The hive, if too valuable to be destroyed, should be thoroughly disinfected before being used again. This may be done by painting all the inside parts with petroleum, and setting it on fire for a moment or two, when, if a wet sack be thrown over it, the fire can be extinguished, and the wood be scraped clean. A painter's blow lamp may be used to scorch the wood. Afterwards, the inside should be well painted over with a strong disinfectant (Recipe 363), and should be left in the open air until the smell of the disinfectant has disappeared.

356. Treatment by Artificial Swarming.—When the owner is disposed to direct his energies to the cure of the disease in its advanced stages, he should proceed by the method of artificial swarming (222). Prepare a skep, with a feed hole on top (318), and place it on the stand of the infected hive, with a hiving board (233) in front. Or, better still, instead of a skep, procure a lidless box (Fig. 110); let into two opposite sides two laths 1½" wide, and 3" or 4" apart, nailing them securely, and attach to the four edges of the laths four slips of foundation running from end to end and not more than ½" deep. A flight hole must be bored just over the bottom in one side. The box illustrated has been successfully used by the author for the purpose. The slips of foundation were attached to the laths by small tacks, and from them the bees built a considerable quantity of comb. Invert such a box on the stand of the infected hive, raised a little in front, and with a hiving board (233) in position. Subdue the bees: close the doors: take out the frames one by one: don't shake, but brush, the bees off into a large sized box, or into a skep, into which brush every bee remaining in the hive. Return the infected combs to the hive, and, for the present, cover up carefully from marauding bees. Now, hive the bees by throwing them down upon the hiving board in front of the prepared box (Fig. 110), or skep, already placed on their old stand. The box, after the bees

Fig. 119.
BOX PREPARED FOR DISEASED BEES.

cluster in it, should be turned up, and a piece of coarse, open canvas, having a round 2" hole in the centre for feeding purposes, should be tacked on as a covering. On this, supported by the laths, a feeding stage and bottle will stand. The box must have a temporary roof, or covering from rain or robber bees, but so that air may reach the bees freely through the canvas covering, and through small gimlet-holes bored in the four sides of the box. The bees may then be fed with medicated syrup (Recipe 322). After four days the infected honey carried from the hive will have been used up as food and for comb building, and the diseased bees will have succumbed. Prepare, accordingly, a clean hive, with four, five, or six frames of foundation: place it on the old stand, with a hiving board in front, and shake the imprisoned bees on to the hiving board. Place a cloth, saturated with a solution of formalin **(366)**, under the floor board ventilator, and continue to feed the bees with medicated syrup for a week or ten days. The skep; the box used in the operation; and the fittings of the old hive should be burned: the hive should also be burned, or disinfected, as directed above **(355)**; and the ground around the stand should be turned over. The frames and combs, it will be safer to burn: but, if it be desired to save the wax, the combs may be thoroughly boiled **(280)**, the wax being extracted for household use: the residue and bag should be burned.

357. Requeening Desirable.—As already stated, foul brood is a disease of mature bees, as well as of brood **(349)**. A failure in the treatment recommended, may be due to the existence of bacilli in the organs of the queen; and, in general, requeening should be practised in connection with the other remedies. The introduction of a young, vigorous queen gives a better tone to the colony, and promotes that active resistance to the disease which is so desirable.

358. Infected Honey Dangerous.—Honey taken from infected stocks, though quite safe for the owner's use, should never be fed back to bees, not even if previously boiled. It is certain that the spores of foul brood can be communicated to larvæ in boiled, infected honey, if fed to them, and that spores will survive chemical treatment, and even freezing, and boiling, such as would at once destroy bacilli.

359. Disinfecting Necessary.—Too much emphasis cannot be laid upon the necessity for a thorough disinfecting of the hands and clothing, and of hives, frames, combs, and all appliances which may have been in contact with this terrible disease —**(355.363.364.)** No remedies can, by any possibility, prove effectual unless they include such disinfection. The ground in the immediate neighbourhood of the hives, also, becomes infected by the throwing out of particles of injurious matter, and of diseased larvæ, and by the accumulation there of bees which had perished by the disease. It should be well dug, and the sods should be turned over. Neglect of this detail has, more frequently than many suppose, led to a recrudescence of the disease, after it had been satisfactorily grappled with and overcome in the hives. A recent extensive experiment, which included the transport for several miles of diseased stocks, and their treatment there, showed very excellent results until, some months later, the stocks were returned to their old stands, when the disease immediately broke out again, and with renewed activity, which speedily exterminated the stocks. Evidently, the infection located in the ground about the old stands had not been dealt with, and the spores of the disease, capable of growth in suitable material, and impervious to weather conditions, finding rich soil for their growth in the healthy larvæ of the returned stocks, took full advantage of it.

359b. "American" and "European" Foul Brood.—In 1908-9 Dr. G. F. White and Dr. E. F. Phillips published in the Bulletin of the U.S. Bureau of Entomology, the results of certain investigations into the nature of Bee Diseases. According to

Photo by]   *[J. G. Digges.*
DISEASED STOCK, "J. G. D." VENTILATOR AND "FEDERATION" DUMMY.

these investigations, which, however, are not yet completed. *Bacillus alvei* **(351)** has been discovered in the disease commonly called Black Brood **(341)**, and to this disease, they, accordingly, give the new title of "European Foul Brood," claiming that it is the disease which was investigated by Cheyne and Cheshire in England; while to the more prevalent disease they give the name of "American Foul Brood," stating that they have "conclusively demonstrated" its cause to be *Bacillus* *larvæ*. The differing symptoms, as described, may be here stated:—

"AMERICAN."

Very prevalent.
Larvæ attacked about the time of capping.
Colour, first light chocolate, and later that of roasted coffee.
Cappings sunken and perforated.
Bees do little to clean out.
Matter is ropy and stretches.
Odour of glue.
Scales, very dark brown, strongly adherent.
Seldom attacks drone or queen larvæ.
Infectious.
Cause—*Bacillus larvæ*.

"EUROPEAN."

Not so widespread.
Larvæ earlier attacked, a small percentage capped.
Colour, first a yellow spot, then all brown and almost black.
Cappings sunken and perforated.
Bees clean out some dried scales.
Mass does not stretch out.
Odour very slight.
Scales irregular, not strongly adherent.
Attacks drone and queen larvæ.
Much more infectious.
Cause—*Bacillus pluton*

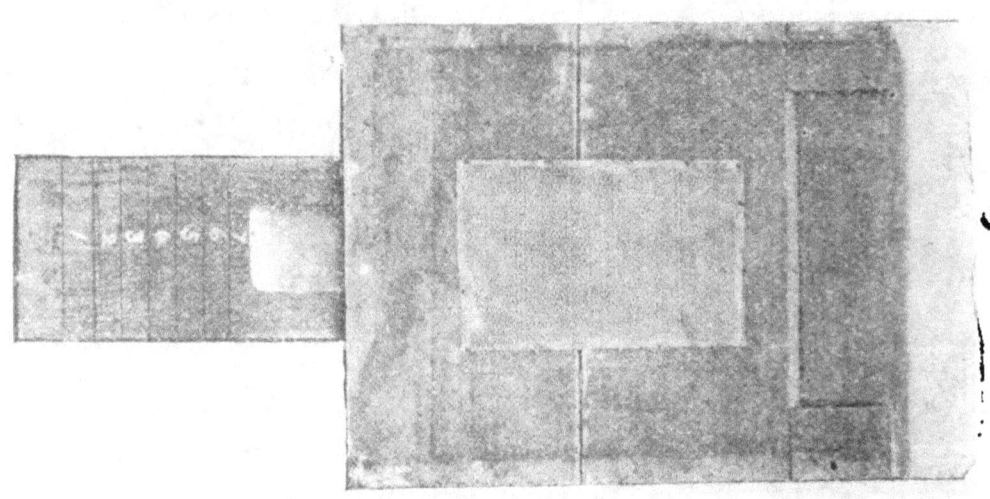

Fig. 111.
"J. G. D." VENTILATOR.

360. "Isle of Wight Disease."—This disease, of which the cause and cure have still to be discovered, made its first appearance in the Isle of Wight in 1904, and was described in the *Irish Bee Journal* (1906) by Mr. H. M. Cooper, Hon. Secretary of the local Beekeepers' Association, who said that the symptoms of the disease were exactly as described in the Irish Bee Guide under the heading of "Paralysis," **(333-5).** At that time ninety per cent. of the stocks in the Island had perished.—

"In some cases several hundreds of bees are to be seen on the ground near the hive, often crawling rapidly, but quite unable to fly, their abdomens greatly distended and containing a large amount of ropy, yellowish-brown matter. The stocks affected rapidly dwindle, and usually succumb in about a month or six weeks, leaving their stores, and often a quantity of brood. The queen appears to keep healthy and survives to the last. Although re-queening and other remedies have often been tried, the results have always been fatal."—*Irish Bee Journal*, June, 1906.

In 1907, the Board of Agriculture and Fisheries deputed Mr. A. D. Imms, B.A., M.Sc., to inquire into the nature and cause of the disease. His report was published in the Journal of the Board, June, 1907, and a further report, by Dr. Walter Malden, appeared in the Journal, February, 1909. Mr. Imms stated that the earliest noticeable symptom of the disease is the inability of the affected bees to fly more than a few yards without alighting. At a later stage the flight extends to a few feet only from the hive, the bees dropping to the ground, and crawling up grass stems or hive supports, and dying soon after. A badly diseased bee crawls with its abdomen dragging on the ground and distended beyond normal proportions. The

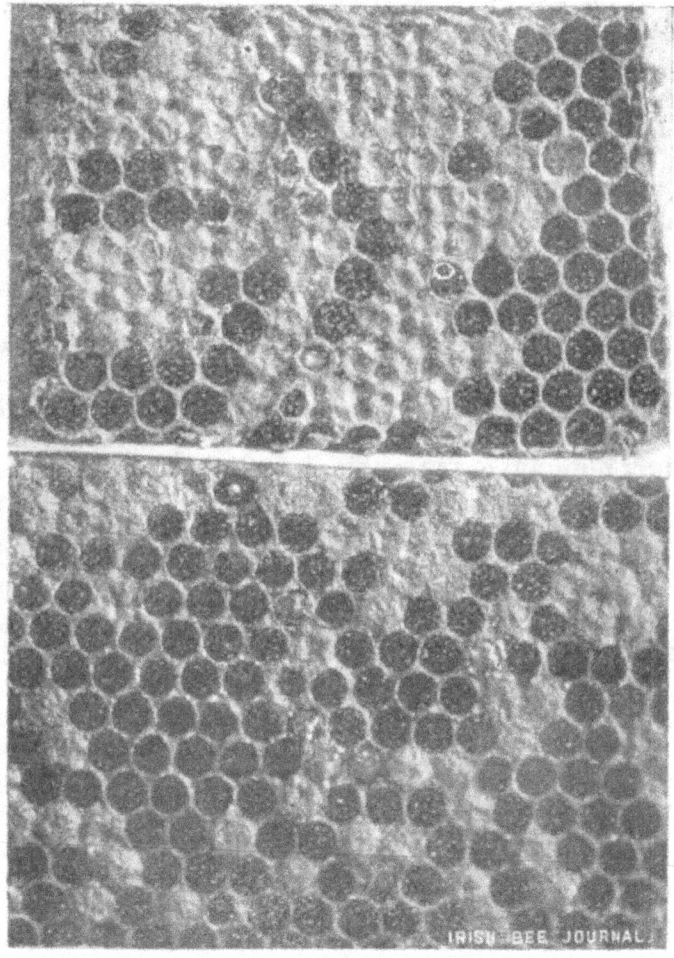

Photo by American Photo Co., Croydon.

Fig. 112.

BEE DISEASE IN THE ISLE OF WIGHT.

Upper comb, from centre of brood nest of diseased hive. Lower comb, from outside of brood nest of same hive, showing young bees in act of emerging from cells.

trembling motion associated with paralysis (334) was not observed by Mr. Imms. The disease is confined to adult bees, and does not appear to affect the brood. The distension referred to is caused by yellowish-brown material filling the colon and containing an enormous number of pollen grains. Dr. Malden found plague-like bacilli in the chyle stomach of diseased bees, and regarded these as the cause of the disease, but he had not fully established their relationship, as he had been unable to demonstrate them in every case. In 1912, the Board of Agriculture published a report by G. S. Graham-Smith, M.D., H. B. Fantham, D.Sc., Annie Porter, D.Sc., G. W. Bullamore, F.R.M.S., and W. Malden, M.D., which described *Nosema apis* as the agent responsible for most cases in which the symptoms of the disease had been noticed, and cited numerous instances in which bees had died after having been fed with honey contaminated with *Nosema apis*, or from an infected hive, or after having been in contact with infected bees: but this report did not claim to be final; it confessed that many problems remained unsolved. In 1916, Dr. Rennie, D.Sc., and John Anderson, B.Sc., reported on the disease to the Royal Physical Society of Edinburgh, declaring themselves unable to recognize any causal relation between the presence of *Nosema apis* and the disease: they found that deliberate infection with *Nosema apis* did not produce the disease, that the disease occurs where the parasite cannot be found, and is not necessarily conveyed by feeding on contaminated stores, or by mere contact with contaminated hives, or combs. In 1920, a report was made of the result of investigations carried out by Dr. Rennie, Mr. Bruce White, and Miss Harvey, of the University of Aberdeen. A discovery had been made of the existence of a type of parasitism in bees, which had been hitherto unknown—an extremely diminutive mite, which invades the respiratory system, and belongs to the genus *Tarsonemus*. This mite (*Acarus*), bred within the bee, is confined to a limited, but important, region of the breathing system. The mites block the air tubes, cut off the air supply from surrounding organs, seem to feed on the blood of the bee, and possibly affect the blood with a specific virus, and, by the investigators, are believed to be the cause of what has hitherto been known as "Isle of Wight Disease," for which name it is proposed to substitute the title "Acarine Disease." The mite has been named *Tarsonemus woodi*. No remedy has been discovered by the investigators; and until such shall have been found, it is recommended that affected stocks be destroyed. Various drugs have been advertized as preventives, or cures, but no satisfactory evidence, either of prevention or cure, by any drug, is forthcoming.

361. Differential Diagnosis.—

	Chilled Brood (336)	Black Brood (341)	Pickled Brood (345)	Foul Brood (349)
CAUSE	Exposure to cold	*Bacillus pluton*	*Aspergillus pollinis*	*Bacillus larvæ*
SYMPTOMS.—CAPPINGS	..	Indented	..	Dark: Indented: Punctured
LARVÆ—Position	Normal	Irregular	Lies on back	Horizontal
„ Colour	Grey to Black	Black	Yellow to Black	Brown
„ Consistency	..	Jelly-like: No ropiness	Watery	Ropy: Elastic
„ Smell	None	Sour	None	Gluey

RECIPES.

Medicated Syrup.— See Recipes 321-322. Page 183.
  „ **Candy—**   „   323-324. „ 184.
Naphthol Beta Solution.— „   325. „ 184b.

362. Carbolic Solution, for Subduing Bees (127).—

 Calvert's No. 5 Carbolic Acid ... ... 1 part.
 Water ... ... 10 parts.

Shake the bottle. Thoroughly damp the cloth, and keep it in a tin box.

363. Solution, for Disinfecting Hives (355).—

 Calvert's No. 5 Carbolic Acid .. .. 1 part.
 Water .. .. 2 parts, or
 Izal .. .. 1 teaspoonful.
 Water .. .. 1 quart.

Paint the hive thoroughly with the solution, and set it in the open air until the smell disappears.

364. Solution, for Disinfecting Clothing, etc. (359).—

 Calvert's No. 5 Carbolic Acid .. .. 1 part.
 Water .. .. 15 parts, or
 Izal .. .. 1 teaspoonful.
 Water .. .. 1 quart.

365. Solution, for injecting into Diseased Cells (354).—

 Formalin .. .. 1 part.
 Water .. .. 4 parts, or
 Izal .. .. 1 teaspoonful
 Water .. .. 1 quart.

366. Solution, for use under Combs (354).—

 Formalin .. .. 1 part.
 Water .. .. 14 parts, or
 Izal .. .. 1 teaspoonful.
 Water .. .. 1 quart.

CHAPTER XXX.

ENEMIES OF BEES.

367. Enemies.—Bees, like every other living thing, have their natural enemies; and, in some countries, it is very necessary to protect them from a variety of foes. In this country, however, all that is required is to keep the stocks strong enough to protect themselves, and to give them hives that do not offer special facilities to the attacks of dangerous intruders.

368. Ants.—These insects (61) sometimes make their nests about the hives, and give a little trouble. But healthy bees are very well able to cope with them. Naphthaline (352) in the hives, and among the quilts, discourages ants. To stand the hives in saucers of water, or tar, will keep out these insects. If they become very troublesome, the nests should be destroyed by making a hole a foot deep through the centre of the nest with a pointed stick, and two or three similar holes around it, when, ½-oz. of bi-sulphide of carbon (355) may be poured into each hole, and the clay may then be closed in. Bi-sulphide of carbon is highly inflammable, and must not be brought near fire or lamp.

369. Birds.—Sparrows, Starlings, Chaffinches, Blue Tits, and even Swallows, occasionally prey upon bees. In hard winters, birds may sometimes be seen on the alighting boards, picking up venturous bees. It is not reasonable to declare war upon, and to destroy these beautiful things for obeying their instincts in search of food, when frost has dried up the earth, and berries are no longer in the hedgerows. All that is necessary is to arrange a yard of old herring netting in front of the hive; thus, at the expense of a penny, protecting each colony until the opening of spring shall offer other provender to the songsters of the woods.

370.—Earwigs.—Earwigs seldom enter the hive proper, but are often found behind dummies; under the shoulders of frames; and among the quilts. They do no harm. Naphthaline (352) will drive them off. Standing the hive in saucers of water, or tar, will prevent their entrance.

371. Mice.—When winter drives the bees away from the entrance, mice will creep into hives and make their nests in the warmth, if the space at the doors is more than ¼" high

They eat honey and chilled bees, and set up a stench which is so highly objectionable to bees that colonies will often forsake such hives in the spring; and, if returned, will refuse to remain. Bees have been known to completely cover up a dead mouse with propolis, in order to suppress the smell; and they will refuse to occupy supers-that have been visited by mice when carelessly stored away in winter. Entrances that are too high should be reduced to ⅜", and all hive fittings, foundation, etc., should be protected from the visits of mice.

372. Parasites.—The Blind Louse (*Braula cœca*)—a red louse, sometimes found upon the bodies of queens and workers. These generally are more numerous upon queens, and are very worrying. More common in warm climates, they have been known to destroy whole colonies in Italy, where it is not uncommon for stocks, affected in this way, to forsake their hives. The lice may sometimes be picked off the bodies of queens. A little tobacco smoke will cause them to drop on to the floor board, when they can be swept into a vessel and destroyed. The Pollen Mite—This insect is often found in pollen cells in weak colonies. Not actually harmful to the bees, strong stocks quickly clear them out of their hives. The Maggot (*Stylops*)—This maggot is commonly found in the abdomens of *Andrenæ*, and other wild bees. It has not been reported as appearing in the honey bee.

373. Wasps.—In autumn, wasps sometimes struggle hard to gain access to the honey in hives; and, as these insects are both stronger and more active than bees, they can do a good deal of mischief in weak colonies. When their attacks become troublesome, a "dark passage" may be constructed on the alighting board, as advised for robber bees **(310)**. Bottles, with a little beer, or sugar and water, if left beside the alighting boards, will attract and catch these enemies. Their nests should be destroyed wherever found; and queen wasps should get short shrift. **To destroy wasps' nests, place small pieces of cyanide of potassium at the entrance of each nest. Never interfere with a wasps' nest during the day: mark it with a piece of stick and white paper, and do the killing by lantern light.**

374. Wax Moth.—The Wax Moth (*Galleria cereana*) is about ¾" long. These moths breed two or three times in one year. They are very fleet in their movements—"the most nimble-footed creatures that I know."—*Reaumur*. They may sometimes be seen flying in front of a hive on a cloudy afternoon in summer. At night, if she can gain admission, the female deposits eggs in the hive. The worms from these eggs devour wax, brood, pollen, and the cast-off skins of bee larvæ, during from fourteen to twenty days, according to the temperature. They spin around their bodies white, silken cases, and further fortify themselves with a coat of wax and their own excrement. They expose only their heads and necks, and these are so strongly helmeted with scales as to be impenetrable to stings. They perforate the combs, and cover them with webs, cocoons, and excrements, speedily working ruin in the hive, and emerging as perfect winged moths. Bees seem to realise fully the danger of admitting the wax moth; and, unless the colony is weak or queenless, the moth will stand but a poor chance of getting into the hive. But, bee-keepers often introduce the mischief to their colonies; for, combs out of use when left lying about, attract the moths, and become fruitful sources of danger when given to the stocks. When the danger threatens, weak colonies, if not united to strong colonies, should be confined to the combs which they can cover and defend; for, if the outer combs be left vacant, the moth, on entering, will be able to work her mischief unhindered. The larvæ of wax moth should be destroyed when found; and comb infested by this enemy should be removed and the wax extracted. Combs that are not beyond saving may be placed in hive bodies, or super boxes, piled up on an empty box in which some ounces of sulphur have been placed and kindled. The receptacles being properly covered, the sulphur fumes will ascend and will kill the moths and grubs. This treatment should be renewed after a few weeks.

Fig. 113.
WATER FOUNTAIN.

CHAPTER XXXI.

WINTERING.

375. Successful Wintering.—So much depends upon the successful wintering of bees, some general advice may, perhaps, be usefully given here. The main points to be attended to are—(1) To winter only strong stocks. (2) To provide a sufficient quantity of wholesome food. (3) To keep the bees as quiet as possible. (4) To supply sufficient ventilation. (5) To avoid damp, and the ill-effects of storms.

376. Winter only Strong Stocks.—Small stocks of bees consume more food, proportionally, than do strong stocks, and are seldom profitable in the following year. Frequently such stocks die out altogether before the spring opens, from inability to keep up the necessary heat of their cluster, and from excessive consumption of food, leading to dysentery **(328)**. Stocks that do not cover at least six frames in the middle of September, should be either strengthened by the addition of healthy, driven bees **(250),** or should be united to each other, or to stronger stocks.

377. Provide a Sufficient Quantity of Wholesome Food.—This has been dealt with under the heading of "Feeding" **(315).** The bees cluster on the empty parts of the combs, just below the honey, the head of each bee under the abdomen of the bee above her; and the food is passed down from one to another until, during a warm hour on some sunny day, the lower bees find opportunity to move up to the food. As the bees on the outside of the mass become chilled, they pass into the warmth of the cluster. But, when the food in the immediate vicinity of the cluster is consumed, the bees, in very cold weather, are unable to move to distant combs, and will often starve to death in the midst of plenty. Therefore, the food required should be given rapidly, towards the middle of September, and the combs with sealed food should be moved to the centre of the hive where the clustering bees may have access to them. Candy, if given, should be placed right over the cluster; and, when candy is not supplied there, "winter passages" should be provided, to permit the bees to pass from comb to comb without having to go under or around the frames in cold weather, when many of them would become chilled, and the remainder, refusing to leave the warmer portion of the hive, would perish from hunger. Two pieces of stick, ¾" thick, laid across the frames, say 1" apart, provide a winter passage under the sheet. Entrances should be sheltered from direct sun rays while snow is on the ground, lest bees, attracted by the light and heat, should fall on the snow and die; and lest those within, encouraged to break up the cluster, should continue in a state of activity throughout the winter, consuming extra food, exhausting their vitality, and, probably, falling victims to disease. Towards the end of February, or the beginning of March, if the weather permit, a corner of the sheet may be raised for a moment, when, if it is seen that food is required, a cake of flour candy (324) should be given at once.

378. Keep the Bees as Quiet as Possible.—Sudden changes of temperature lead to increased activity in the cluster, and this means increased consumption of food, with the frequent result of filling the intestines with digested food, which, the bees being unwilling to void it in the hive (11), promotes dysentery (328). Therefore, unnecessary, empty combs should be removed, and the nest should be reduced to the size required by the bees, the dummy being moved up for the purpose. Warm coverings should be placed over the frames. A section crate, with a piece of stuff tacked underneath, and filled with cork dust, dry chaff, or torn paper, may be set upon the quilts (96). With hives constructed for the purpose, the riser (87) may be inverted over the body box, thus supplying additional walls; and the porch may be transferred from the body box to it. (Fig. 116, a.).

Fig. 114.
THATCHED SKEP.

379. Supply Sufficient Ventilation.—When danger of robbing is over, the bees having ceased to fly freely, the doors of all hives should be opened to a space of about six inches; and, frequently during the winter, a crooked wire should be used to draw out any dead bees which, accumulating near the entrance, might prevent ventilation, and lead to the smothering of the living bees within.

WINTERING.

380. Avoid the ill-effects of Damp and Storms.—Care should be taken to make all hive roofs thoroughly waterproof, because, damp entering is very unhealthy. Damaged roofs should be repaired and well painted before the winter sets in; and, where necessary, waterproof covers should be put on in anticipation of severe rain and snow. Elvery's waterproof cover can be put on and taken off in a minute or two and gives complete protection. A defective roof can be made rainproof by giving it a coat of thick paint, then laying on, while the paint is wet, a piece of canvas or calico to cover the roof top. This material should have a good coat of paint, and a second coat when the first has dried. Skeps require special attention in this respect. They should have a thick covering of straw, tied securely at the top, and held to the skep by hoops (Fig. 114). Storms sometimes make havoc of stocks by upsetting, or unroofing the hives. A stake should be driven into the ground close to the hive, and a rope, carrying a heavy stone, or a couple of bricks, and tied to the stake at one end, should be passed over the hive as a protection against storms (Fig. 115). Snow should be brushed off the hive roofs before it melts.

Fig 115.
HIVE SECURED AGAINST STORMS.

a
Photo by]

b
[*J. G. Digges.*

Fig. 116.
HIVES PREPARED FOR (*a*) WINTER AND (*b*) SUMMER.

CHAPTER XXXII.

WORK FOR THE MONTH.

381. January.—Towards the end of the month, if stores are required, give a cake of flour candy **(324)** under the quilts. Disturb the bees as little as possible. Examine under the roofs for damp, and make repairs where necessary. Remove dead bees from entrances **(379)**. Attend to instructions under the head of "Wintering" **(375).**

382. February.—Feed with flour candy **(324)** where necessary. Remove dead bees from entrances **(379)**. Replace damp quilts by dry ones. Tidy up the apiary. Overhaul hives not in use; and clean and disinfect them **(355).**

383. March.—Feed with flour candy where necessary **(324)**. In a case of impending starvation, give a quart or two of thick, warm syrup (Recipe **322**) as fast as the bees will take it. Supply artificial pollen **(192)** and water **(319)**. Uncap some honey in combs, every few days; and, about the end of the month, begin stimulative feeding **(313)**. Reduce entrances to one bee space as a protection against robbing **(310)**. Unite queenless stocks to others having queens **(247)**. Keep a look out for signs of disease **(327)**. Order sections, foundation, hives, and other appliances for the coming season. Prepare sections **(257)** and wired frames **(261)**. Have all vacant hives cleaned and ready for "spring cleaning" next month **(252)**. Sow seeds of bee flowers (Chapter XXXIV.).

384. April.—Attend to spring cleaning, transferring stocks to clean hives **(252)**. Add warm, dry wraps where required. Keep a look out daily for robbing **(307)**. Examine all stocks for signs of queenlessness **(283)** and disease **(327)**. Begin to build up stocks, so that they may be at full strength not later than June 1st **(311)**. Unite weak stocks, saving the better queen **(246)**. Continue stimulative feeding **(313)**. Spread brood, using all the precautions recommended **(193)**. During manipulations, guard against robbing **(308)** and chilled brood **(338)**. Fix foundation in sections **(258)** and frames **(263)**. Prepare for queen rearing **(286)**. Destroy queen wasps where found **(373).**

385. May.—Continue stimulative feeding **(313)**. Spread the brood **(193)**. Supply water if required **(319)**. Open doors to full width when honey comes in plentifully and danger of robbing is over. Prepare and furnish sections, crates, super boxes, and frames of wired foundation **(255)**. Double strong stocks **(270)**. Watch for signs of honey flow **(265)** and give supers in good time. Provide against swarming by giving room and ventilation **(218)**. Prepare hives for swarms. Proceed with queen rearing **(286)**. Form nuclei **(290)**.

386. June.—Supply supers as soon as honey flow begins, and add fresh supers as required **(269)**. Make artificial swarms where desired **(222)**. Attend to nuclei **(290)**. Give ventilation **(218)**. Extract honey from combs in body box, and return extracted combs to centre of brood nest **(217)**.

387. July.—See that ventilation is sufficient in hot days **(218)**. Add supers as required **(269)**. Supply young queens from nuclei **(295)**. Extract from combs in body box **(217)** as recommended in June.

388. August.—Extract honey, as directed for June and July **(217)**. Transport stocks to heather **(156-158)**. Remove supers at end of honey flow **(272)**. Begin stimulative feeding **(314)**. Guard against robbing **(308)**. Examine roofs, after hot weather **(88)**. Increase stocks by driven bees **(159)**.

389. September.—Unite weak stocks **(246)**. Strengthen stocks with driven bees **(250)**. Begin autumn feeding about September 15th, and complete it by September 30th **(315)**. Reduce entrances to prevent robbing **(310)**.

390. October.—Reduce brood nest to space required **(378)**. Give candy if required **(316)**. Provide winter passages **(377)**. Give warm wraps. Attend to instructions for wintering **(375)**. Clean and store appliances. Protect hives against storms, rain and snow **(380)**. Plant crocus. and other pollen and honey yielding bulbs (Chapter XXXIV.).

391. November.—Attend to winter feeding, if required **(316)**. Open entrances to six inches, if danger of robbing is over. Make any necessary changes in the apiary **(156.283)**. Attend to instructions for wintering **(375)**.

392. December.—Remove snow from roofs before it thaws. Shield entrances from direct sunshine, while snow is on the ground **(377)**. Remove dead bees from entrances **(379)**. Attend to instructions for wintering **(375)**.

[*Advice upon all matters connected with beekeeping may be had on application to the Author, addressed—" Editor,* BEE PUBLICATIONS, *Lough Rynn, R.S.O., Co. Leitrim, Ireland."* **(161)**. *Queries are replied to by telegraph or post, or in the columns of the Journal, or Gazette.—If required by telegraph,* 1/- *should be forwarded ; if per post, a stamped, addressed envelope should be enclosed for reply. Telegrams : "Digges, Mohill."*]

CHAPTER XXXIII.

EXHIBITING AND JUDGING BEE PRODUCTS.

393. Points to be Aimed at.—The chief features of excellence which are looked for by judges of Bee Products at the leading shows may be summarized as follows, the marks attached to each being those approved and recommended by the Irish Bee-keepers' Association, March, 1910:—

SECTIONS.—*Completeness of filling*, including weight and freedom from popholes and unsealed cells, 25; *condition*, including uniformity of cappings, flatness of surface, and freedom from "travel stain," propolis, "weeping," bruising, and other disfigurements, 20; *flavour and aroma*, 20; *colour* of cappings, 10; *general appearance* of the exhibit, including squareness and cleanness of the wood and glass, glazing, suitability of paper decorations—if any (the overlap of paper not to exceed ⅜"), method of staging—if any, and general attractiveness, 25. Total marks, 100.

EXTRACTED HONEY (Liquid).—*Colour*, which, in classes for "Light" Honey, may range from clear to a pale straw tint, "Medium" Honey from light to dark, Heather Honey to dark brown, 10; *density, or thickness*, 30; *flavour and aroma*, 25; *condition*, including clearness and freedom from froth, air bubbles, suspended matter, and granulation, 20; *general appearance* of the exhibit, including quality and make of bottle, or jar, safety from leakage, neatness of label, and general attractiveness, 15. Total marks, 100.

EXTRACTED HONEY (Granulated).—*Colour*, which may range from white to amber, and, in the case of Heather Honey, to dark brown, 10; *condition*, including regularity and completeness of granulation and fineness of grain, 35; *flavour and aroma*, 25; *general appearance* of the exhibit, including quality and make of bottle, or jar, neatness of label, and general attractiveness, 15. Total marks, 85.

SUPERS OF HONEY, exhibited as removed from the hive, without re-arrangement or cleaning.—*Preliminary preparation, and condition as exhibited*, including squareness of sections, suitability of separators, follower, and spring, evenness of comb, and freedom from "travel stain," propolis, and other disfigurements of super, sections, or frames, and comb, 20; *weight of contents*, 20; *uniformity and colour* of cappings, 10. Total marks, 50.

BEESWAX.—*Colour*, ranging from lemon to pale amber, 20; *freshness, cleanness and purity*, including absence of dross, 20; *aroma*, 10; *texture*, including freedom from brittleness, 10. Total marks, 60.

MEAD.—*Flavour*, 10; *clearness and brilliancy*, 10; *attractiveness* of bottle and label, 10. Total marks, 30.

VINEGAR.—*Flavour*, 10; *clearness and brilliancy*, 10; *attractiveness* of bottle and label, 10. Total marks, 30.

394. Early Exhibition Sections.—As it is necessary to have the cappings of exhibition sections uniform in colour and perfectly free from "travel stain," the sections must be finished as quickly as possible, and must be removed from the hive as soon as they have been finished. If an exhibit of the current season be needed for an early show, select more than the required number of good, clean sections well filled with comb from the previous season, and, with the opening of the first honey flow, place these, over an excluder, on the strongest stock that gives the whitest cappings **(46-49)**, and wrap them up as warmly as possible at the sides, ends, and on top of the crate. Defer as long as you safely can the addition of an extra crate, and if the exhibition lot is not fit for removal when a second crate must be given in order to prevent swarming, leave the first crate undisturbed, and give the second crate on top. But, assuming that the first crate was set upon nine frames only—which is common enough in the early season, the addition of a second crate may be postponed by giving an extra frame, or frames, as required, in the brood nest, and this will help to prevent a slackening of work in the crate. Should a swarm issue, hive it on the old stand in a new hive fitted with only six or seven frames of foundation; add a half inch board **(266)** behind the dummy to prevent the escape of bees there; place the excluder, and the exhibition crate with its bees, on top to be finished by the swarm. When completed, remove the crate by means of a super clearer **(274, 275)**, and in no other way, or you will probably have broken cappings to destroy your exhibit **(397)**. Be careful that no cappings become bruised when the sections are being taken from the crate; proceed in this way,—remove the spring, or wedge, from behind the follower **(106,** and Figs. 35 and 36), place a tray, or a clean, flat board upon the crate, carefully turn all over on a table, then loose the crate and lift it off the sections, when the latter may be separated without risk of being damaged.

395. Mid-Season Exhibition Sections.—Sections intended for the Clover, or "Light" Honey exhibition classes, should be worked from the foundation in the current season. Before the flow from clover opens, select a stock that is already doing good work, through an excluder, in its first crate, and that can be relied upon for white cappings **(46-49)**, and when clover begins to yield, give that stock, under its first crate, a crate of clean, new sections, carefully folded, and furnished with full sheets of worker-comb foundation stopping an eighth of an inch from the bottom of each section. Do not disturb the stock again for a week. If then the first crate proves to be ready for removal, remove it with the aid of a super clearer **(274-275)**, without in any way disturbing the exhibition crate, and give a new crate on top. On the question of giving the bees more room before the completion of the exhibition crate, careful judgment must be exercised. If more frames should be required, or another crate, or should the stock swarm, and for the removal of the exhibition crate and the separation of the sections, the instructions given in the preceding paragraph **(394)** should be followed.

396. Heather Sections for Exhibition.—If it is intended to enter competition in the classes for Heather, or "Dark," Honey Sections, select more than the required number of incomplete, but well-built sections from the earlier season's crates; extract the honey, and give the sections, wet from the extractor and over an excluder, to your best stock, placing the choicest specimens in the central positions in the crate. Not more than one or two crates should be given to that stock, and such exhibition crate, or crates, should be wrapped as warmly as possible round the sides and ends, and on the top. Should the bees appear to be unable to use to advantage all their available space before the exhibition sections are completed, do not hesitate to reduce the space, either by removing sections, or by withdrawing frames, so that the required sections may receive from the bees all the attention necessary. For the subsequent details, as to removing the sections from the hive and from the crate, the preceding instructions should be followed **(394)**.

397. Selecting Exhibition Sections.—The most unselfish and unbiased skill, care, and judgment must now be brought to bear upon the delicate task of selecting the sections for exhibition, for it is at this point that so many exhibitors come to grief. Assume that the sections belong to your bitterest enemy—if you have such, and that you are the appointed judge, bound in honour to give them the most critical examination and to discover the slightest defect. During the selection, keep in mind the following essential considerations:—Sections for exhibition should be filled to the wood on all sides and completely sealed. Sections 2" wide and so filled and completed will weigh 18 oz. or 19 oz. gross, including the wood, or 17 oz. or 18 oz. net. The wood of a section weighs 1 oz.; therefore, a section of honey which does not turn the scales at 17 oz. is not entitled to full marks for weight, and is, accordingly, unsuitable for competition. Bulging, or any unevenness, of the comb is a fault which must be excluded from the show bench; the surface of the comb should be quite flat corresponding with the cut-away sides of bee-way sections, or in the case of no-bee-way sections, about ⅛" short of the edges of the wood on both sides. "Travel stain," which is the result of leaving sections too long in the hive, detracts from the appearance and sacrifices marks in competition. Propolis on the wood must be removed. Punctured cappings (394) are to be avoided; they are generally the result of a defective method of clearing the bees from supers,—much smoke, or excessive use of carbolic or other intimidating medium, having the effect of driving the bees to gorge at the cells, and thus the appearance of the contents is injured and their value reduced: this defect may be obviated by the proper use of super clearers (274, 275, 394). "Weeping" describes the condition of a comb that has been stored in a cold, damp place, the honey, with its absorbed moisture, exuding through the cappings in minute drops (302): to state the cause is to describe the necessary precautions to be taken; no weeping sections have any chance with a moderately competent judge (303).

398. Preparing Exhibition Sections.—Having made your selection, prepare the sections for display on the show bench (303). With a cabinetmaker's scraper, a piece of glass, or a blunt penknife, scrape the wood of the sections thoroughly, avoiding any injury to the comb, and finish off with fine sandpaper. No matter what covering or ornamentation it is intended to subsequently employ, this cleaning of the wood should invariably be attended to because it is right and seemly in itself, and because any experienced judge will look for it. The sections may then be glazed, as described (304), neatness, taste, and the most scrupulous cleanliness being essential here: the overlap of paper, or of lace paper, should not exceed

Fig. 117.
GLAZED EXHIBITION CASE.

¼", and all tinsel and gaudy colours should be rigorously excluded. Sections may be shown in special boxes made and sold for the purpose **(304, and Fig. 102, page 171)**, or in exhibition cases (Fig. 117), or failing any of the foregoing, they may be wrapped in wax paper and tied with narrow ribbon. All extravagant and fantastic designs of decoration should be avoided. The exhibit should be carefully packed ready for despatch, and should be kept in a warm place, meanwhile, to avoid "weeping." Where it is possible to do so, exhibitors should stage their own exhibits, leaving them in the best order and condition for the judge.

399. Extracted Clover, or "Light," Honey for Exhibition.—To secure suitable specimens of Clover, or "Light," Honey, for extraction, it is desirable to have on hands a supply of frames of good, clean combs, absolutely free from honey and pollen; they should have had their honey extracted, and have been given back to the bees, over a super clearer, to be cleaned **(274)**, and should then have been carefully wrapped up and stored until required. Immediately upon the clover coming into bloom, the frames should be given, in a super box and over an excluder, to a strong stock, and should be removed, whether finished or not, so soon as the flow from clover ceases.

400. Extracting and Preparing Clover, or "Light," Honey for Exhibition.—Extracting may be carried out according to the instructions already given **(276-278)**. The extractor **(134)**, strainer **(136)**, and ripener **(136)** must be as clean as it is possible to make them, and nothing must be permitted to add either flavour or colour to the honey after its removal from the hive. Density, which is an essential qualification, cannot be secured to the full extent in honey extracted from unsealed cells, because such honey has not been thoroughly ripened, and for show purposes it will not do to ripen it artificially. If, therefore, the combs to be dealt with contain the least quantity of unsealed honey, that honey must first be extracted and stored away, and then the remainder of the combs may be uncapped and their contents may be extracted for exhibition; or, as an alternative, such combs may be uncapped, and revolved in the extractor at a speed only sufficient to throw out the unripe honey, which must be drawn off, the combs being then revolved at the speed necessary to extract the ripe honey required. One week after extraction **(276)** and straining **(277)**, the ripe honey may be run off from the bottom of the ripener, and should be kept in bulk, in an air tight tin and in a warm place. Three or four days from the date of the show at which the exhibit is to be made, the tin of honey should be set in a vessel of hot water until the honey reaches 80° Fahr., when it may be run into the selected jars. The jars must be left in a warm place, covered from dust, until all air bubbles, or scum, in the honey shall have risen to the top, when the bubbles, or scum, must be carefully skimmed off, and, if necessary, an addition of ripened and skimmed honey should be added to bring the contents of each jar up to 16 oz. Uniformity being necessary, with respect to flavour, colour, and density, if there is any difference in the exhibits, the quantity required for the jars should be mixed in one vessel beforehand. Care must be taken to exclude any honey that may have been tainted with honey dew (61), because such an admixture would utterly spoil the colour and flavour of the exhibit. The jars for exhibition purposes must be carefully selected, of clear, flawless glass, and, preferably, with screw caps fitted with cork wads (306). For each jar, cut a circular piece of wax paper the same size as the cork wad; put this on the mouth of the jar, set the cork wad upon it, and screw the cap tightly home. A neat label should be added (306).

401. Extracted Heather, or "Dark," Honey for Exhibition.—As in the case of Heather sections (396), built out combs, wet from the extractor, should be used over an excluder. Owing to the difficulty of removing heather honey from the combs in an ordinary extractor, if the extractor is to be used the combs to be employed should be tough and strong, and preferably drone combs. If the honey is to be extracted by means of the Honey Press (137), or by melting (402), fresh, virgin combs will serve best.

402. Extracting and Preparing Heather, or "Dark," Honey for Exhibition.—If it is intended to remove the heather honey by means of an extractor, everything required should be in readiness, in a warm room, to extract the honey hot from the hives, for if it be allowed to cool, extraction will be exceedingly difficult. It will be found more practicable to crush the comb in a Honey Press (137, 276), or to melt the wax. In the former case, the combs should be heated up to 120° Fahr., being placed in the Press as directed (276). If the melting process is to be adopted, the sealed combs should be cut out and placed in a tin vessel, which should then be set in a pot of warm water, with a wire mat, or other suitable device, underneath, to keep the tin about 1" up from the bottom of the pot; the water must then be heated gradually, and the contents of the tin must be stirred frequently until the wax begins to melt, at which point the temperature must be maintained until all the wax has melted, for if the melting point of wax (144°, 62) be exceeded, the flavour of the honey may be spoiled (306). When all the wax has melted, the contents of the tin must be allowed to cool until the wax can be lifted off the top in a cake, after which, without further cooling, the honey may be strained into the selected jars and treated as described above **(400)**.

403. Supers of Honey for Exhibition.—The object of this class is to encourage careful handling of the sections, frames, foundation, crates, and super boxes, and to judge of the capability of the exhibitor through the perfection, or otherwise, of his entire exhibit, his aim being to produce the maximum of good comb and honey with the minimum of propolis, travel stain, popholes, and other detractions. If the crate, or super box, is to be exhibited exactly as taken from the hive, none of the contents may be handled subsequently, nor may any marks or stains be removed. Accordingly, all the more care must be taken with the preliminary details, viz. :—The choice of a stock that may be relied upon to give good work and to finish with white cappings **(46-49)**; the selection of the crate, or super box, only such as are absolutely accurate in all their measurements **(103, 105)** being employed; the folding of sections and the putting together of frames; the insertion of foundation, separators, follower, and spring or wedge; the application of vaseline, or petroleum jelly **(174)**, to minimise propolising; the accurate fitting and evenness of sheet and quilts; the careful wrapping of the crate, or super box, with warm materials, to conserve heat and expedite the work. If these details be properly attended to, the results should be satisfactory, but if any of these details should be neglected, failure will probably follow.

404. Beeswax for Exhibition.—The best results are obtained from cappings and virgin comb. When preparing combs for the extractor, the whitest cappings should be taken off with as little as possible adhering comb, and be set apart for exhibition purposes, and when the honey from them has drained off, they should be left in a vessel of clean rain water for a few days. Hard water, or water containing lime, should never be used in any of the processes adopted for wax-rendering, for it injures the quality of the wax. Upon removal from the water the wax should be dried, kneaded into small balls, and inserted, preferably in a steam wax extractor **(140)**, or, if such an extractor is not available, the wax may be put into a perfectly clean earthenware jar in a moderately hot oven, or in a pot of boiling water on the range. When the wax has melted it should be strained through fine muslin into a bowl of warm water, and allowed to cool slowly, for rapid cooling produces cracks in the wax. When cool, and before it is perfectly cold, the cake of wax should be lifted off the water, and should have all dross and dirt scraped away;

it may then be broken up, re-melted, and poured through muslin into a suitable mould, or moulds, previously wet with clean water, and should be allowed to cool as slowly as possible. The processes of melting, straining, and scraping may be repeated so long as there remain any impurities to be removed, but they should not be carried to the point of injuring the texture of the wax and making it brittle. As elsewhere stated **(280)**, dark wax may have its colour improved by the addition of sulphuric acid (vitriol) to the water in which it is to be melted. When old combs are being dealt with for exhibition purposes, the following method may be adopted with advantage:—Set two vessels of hot water side by side on the range; into one crush as many combs as it will hold, leaving some inches to spare for the swelling of the wax when it boils. As the wax melts, skim it off, as free as possible from dirt, into the second vessel, and discontinue this process when the wax becomes too dirty for the purpose; the first vessel is then to be emptied and cleaned, its wax contents being reserved for further treatment. Now put some boiling water into the empty vessel, and also a large, clean jam crock containing some boiling water; strain the wax from the second vessel, through fine muslin, into the crock; then stir it with a thin piece of wood, and, as you stir, drop a little sulphuric acid, drop by drop, on to the wax; this will improve the colour and will help to remove any impurities that may have escaped the strainer. Now remove the vessel containing the crock and wax to the side of the range, cover it with a lid, and let the cooling be very gradual. When the cake of wax is cool scrape, or cut, from it all impurities. From a number of cakes so prepared select the best specimens, weighing in all a little more than is required, and re-melt these in the crock after having thoroughly cleaned the latter. Damp the inside of your mould with clean water, pour in the melted wax, and set the mould in a pot of hot water where the wax may cool as slowly as possible to avoid cracks. This method, even when applied to combs black with age, has resulted in first prizes at leading shows, where competition was exceptionally keen, the careful skimming of the wax before it had time to become discoloured, and the subsequent processes, having produced cakes of wax of exceptional merit. As an alternative to the former methods—although not one which can be as strongly recommended—the boiling process **(280)** previously described may be adopted, the wax being ladelled off as it rises, to be treated as advised above.

405. Mead for Exhibition.—Mead for the show bench should be well flavoured, full bodied, clear, and, if possible, sparkling. The honey used in its manufacture should be light and well ripened, and throughout the whole process the utmost cleanliness must be secured, not only in the ingredients, but also in every vessel employed. Use 4 lbs. honey to each gallon of water, and allow the honey to dissolve, then put it into a copper, or large boiler, add 1 oz. hops and ½ oz. ginger per gallon, and boil it for one hour, skimming off the scum as it rises. When sufficiently boiled, pour it into a wooden vessel, and when its temperature has reduced to 120° add 1 oz. of brewer's yeast per gallon, mix this well with the liquor, which must then be covered and allowed to stand in the vessel for about eight hours. Next it must be poured into a perfectly clean barrel, and as the contents ferment, the barrel must be filled up with more of the liquor, an extra half-gallon having been prepared for the purpose beyond what the barrel is constructed to hold. When fermentation has ceased, dissolve ¼ oz. of isinglass in a cupful of water, pour it into the barrel, and stir well; this is to clear the liquid. After about six days draw off the liquor into a second perfectly clean barrel, filling the barrel completely, and drive in the bung as tightly as possible. It must stand for at least six months, after which it may be bottled. The bottles must, of course, be perfectly clean, the corks should be new, and they should be fastened with wire, and covered with tinfoil, a neat label being pasted on the side of each bottle.

406. Vinegar for Exhibition.—For the production of a superior exhibit of vinegar, all that is necessary is to use the right ingredients, to study cleanliness in all the processes, and to regulate the temperature with a certain degree of accuracy. Take 1 lb. of good extracted honey, add it to 7 lbs. of fresh clean water in a wooden vessel (or 1 lb. of honey to 5½ pints of water), and stir the mixture thoroughly. Cover the vessel with two thicknesses of fine muslin, and keep it at a temperature of about 80° Fahr. It may be exposed to the sunshine in summer, being brought into a warm kitchen for the night. After about six weeks, if the vinegar proves right to the taste, strain it into another wooden vessel, stir in ¼ oz. of isinglass dissolved in a few ounces of water, and allow it to stand for a fortnight; then bottle it in clear glass bottles, using new corks, which may be covered with tinfoil; put on an attractive label.

407. Judging Bee Products.—No one who accepts appointment as a judge of Bee Products hopes to please and satisfy all the exhibitors; but if he desires to do absolute justice, and to carry out his mission creditably, he will be wise to adopt a fixed scale of marks for the various points, and to rigidly adhere to those marks. By no other method can judging be conducted satisfactorily. In a previous paragraph (393) the points have been described, and attached to each is the scale of marks adopted and recommended by the Irish Beekeepers' Association in 1910. No judge can go far astray who follows the lead thus given. He will require a glass taster—which can be procured for a few pence—a magnifying glass, and a scales with the necessary weights up to 20 oz., which should be supplied by the Show Committee. He should also be provided by the Show Committee with a supply of judge's cards, which should contain, in parallel columns, spaces for the exhibitors' numbers, for the marks to be awarded under each point, for the total marks obtained by each exhibitor, for the maximum marks possible, and for the award, with a space in which the judge's remarks upon any exhibit may be entered opposite the number and marks of that exhibit. As the judge proceeds to examine the sections, he will first enter the maximum marks possible at the head of the columns for points (if this has not been already done), and then he will enter the numbers attached to the exhibits, in vertical order, in the first column on his card; next he will test each exhibit for "completeness of filling, including weight and freedom from popholes and unsealed cells." The exhibits will be weighed, and each exhibit that turns the scale at 17 oz. will be entitled to full marks for weight (say 15), nor will any competent judge award extra marks for weight over 17 oz., no more than he would to 1¼ lbs. of butter exhibited as a 1 lb. roll; if the sections are free from popholes and unsealed cells they will be entitled to full marks (say 10), thus securing the maximum of 25 marks under the first point, and the marks will be entered, under their proper heading, in the second column of the card. The exhibit will next be examined for the other points set forth in paragraph 393, the marks being extended in their proper columns accordingly. Any sections that come short of the requirements will lose marks proportionally. The judge will then proceed with the remaining exhibits, and mark them as they deserve. The points of excellence required in extracted honey, beeswax, mead, and vinegar have already been described (393), with the marks to be assigned. When judging extracted honey for density, or thickness, the jars should be inverted, and the rising of the air bubbles should be accurately timed, the highest marks being awarded to the exhibit in which the air bubbles rise slowest, having regard to the air space in each jar. Weight should be judged by the scales; 1 lb. bottles should contain 16 oz. of honey, short weights being penalised, and extra weights deriving no advantage. "Granulated honey" should be granulated, and not merely thickened. For the judging of wax the magnifying glass will be useful. Supers of frames and crates of sections should be carefully scrutinized for signs of cleaning and of substitution of frames from other supers or of sections from other crates, and, assuming that the conditions laid down **(393)** apply, any exhibit which shows signs of having been improperly manipulated for the show bench, should be disqualified. Mead should be well flavoured and clear, and should be securely corked in glass bottles, bearing suitable labels. Vinegar should show similar qualities, and should be put up in clear glass bottles, well corked and labelled. When all the exhibits in any class shall have been marked for their various points, the judge should tot the marks for each exhibitor in that class, and enter the totals in the column provided for that purpose; above these will appear the total maximum marks obtainable, and the last column will announce the awards—1st, 2nd, 3rd, V.H.C., H.C., C., according to the rules of the particular show. Thus the exhibitor will receive an award according to the total of his marks, and the judge himself will not know the results until he has made his tots; in the margin he will enter any special remarks upon any exhibit, as he may think desirable. Judge's cards, embodying the above details, have been published by, and may be obtained from the office of *BEE PUBLICATIONS*, Lough Rynn, R.S.O., Co. Leitrim. When such cards have been completed, signed by the judge, and placed in position on, or over, the exhibits, the competitors and the general public can see in what respects the several exhibits have been successful, or the reverse, and the show becomes, not only a means of awarding, or gaining, prizes, but also an object lesson in the science and practice of Beekeeping, with educational advantages of great use and importance.

CHAPTER XXXIV.

BEE FLOWERS AND PLANTS.

Spring.—Among the garden flowers which are most useful to bees are those which bloom before the field flowers, and after the Clover and Lime:—of the former, Aconite, Crocus, Hellebore, Scilla, White Rock, and Aubretias, in which bees revel during every sunny hour from January to April; and Limanthes Douglasii, a prime favourite in May. Of trees and shrubs, Pyrus Japonica, Cotoneaster, Box, Sally, Gorse, Willow, Broom, Ribes Rubra, and Gooseberry yield largely in the opening months of the year, and are followed by Sycamore, Hawthorn, and fruit trees, which usher in the honey flow, and usually give bees continual employment until White Clover and Sainfoin begin to yield. Of the foregoing, those which produce honey in quantity, and of a distinct type are:— Sycamore—honey heavy, somewhat green in tint, and lacking in flavour. Hawthorn—honey heavy, amber coloured, flavour and aroma delicious. Fruit trees—honey excellent, in colour and consistency resembling that from Sycamore.

Summer.—The main honey flow, which occurs in summer, is from White Clover and Sainfoin, commencing about the beginning of June and continuing for about a month. Ragweed also yields at this time. Then the Lime carries on the season until the end of July, which terminates the honey flow except in heather districts. White Clover and Sainfoin yield the thinnest and lightest coloured honey, of a most agreeable and delicate flavour. Ragweed, which grows in profusion all the time of White Clover, gives a most disagreeable honey, and often spoils that gathered at the same time from Clover; its honey is rank and coarse like the flower, and has an objectionable aroma. Lime gives a heavier honey than that from Clover, and of a much deeper hue, the cappings of the combs being straw colour. Between the Lime and Heather, Saxifrage, Poppy, Borage, Mignonette, Canterbury Bells, etc., provide good forage for bees. Blackhead (*Centaurea Nigra*) blooms at the same time as heather, and, being a prolific source of nectar, is often preferred by bees. Its honey is thin, of a rich amber colour, and acrid in flavour. Heather honey is quite distinct from any other; its colour is deep, often approaching purple; and it crystallises to an unattractive brown. Its flavour is rich and strong; and its consistency is so thick as to defy the powers of an extractor. Ling Heather (*Erica vulgaris*) (Fig. 118, a) is the most abundant yielder. Its height seldom exceeds one foot; its leaves are tiny green; and its flowers also are small, pale pink, varying to deep purple, or white. Bell Heather (*Erica cinerea*) (Fig. 118, b) is more bushy than the former; its leaves are smaller, and its flowers are a reddish purple. Cross-leaved Heather (*Erica tetralix*) (Fig. 118, c) is short, with small leaves, growing in fours, crossways, up the stem; its flowers grow in clusters of from five to twelve at the top of

Fig. 118.
HEATHER BLOOMS.
a. *Erica vulgaris* (Ling Heather); b. *Erica cinerea* (Bell Heather); c. *Erica tetralix* (Cross-leaved Heather).

the stem; the bells are pale pink in colour, edged with four pointed teeth: this heather flourishes only on damp bog land, and is of little value as a honey producer. In late summer, the Blackberry attracts bees to the hedgerows, and yields large quantities of honey and pollen.

Autumn.—Ivy, which, if left to grow of its own sweet will on walls and trees, blooms profusely in October, is eagerly sought after on sunny days. The honey it yields is very inferior, but it makes a useful addition to winter stores in the hives.

COMPARATIVE LIST OF BEE PLANTS AND FLOWERS.
(Yield:—G, good ; M, medium ; P, poor.)

NAME.	Date		For Pollen	For Honey.
	From	To.		
Aconite,	January	April	P	M
Box	February	March	G	M
Crocus	,,	April	G	P
Dandelion	,,	October	G	P
Hazel	,,	March	G	Nil.
Hellebore	,,	May	G	M
Sedum Major	,,	April	M	G
Snowdrop	,,	March	P	M
Aubrietias	March	July	P	M
Barberry	,,	April	G	P
Cotoneaster	,,	May	P	M
Gorse	,,	June	G	M
Mallow	,,	,,	G	M
Poppies, Single	,,	,,	G	Nil.
Ribes Rubra,	,,	May	M	G
Sallow	,,	,,	G	G
Scilla	,,	,,	P	G
Violet, Sweet	,,	,,	P	G
Wallflower,	,,	,,	M	G
White Rock	,,	June	P	M
Willow	,,	May	G	G
Cherry	April	,,	P	M
Gooseberry	,,	,,	M	M
Pyrus Japonica	,,	June	M	M
Pear	,,	May	M	M
Plum	,,	June	M	M
Sycamore	,,	May	M	G
Apple	May	June	M	G
Bird Cherry	,,	,,	M	G
Broom	,,	,,	G	P
Cabbage,	,,	July	G	M
Hawthorn	,,	June	M	G
Holly	,,	July	M	G
Limauthes Douglasii	,,	June	M	G
Mignonette	,,	November	G	G
Ragweed	,,	August	G	G
Raspberry	,,	June	M	M
Strawberry	,,	,,	G	M
White Clover	,,	July	M	G
Bokhara Clover	June	September	M	G
Borage	,,	November	M	G
Buckwheat	,,	July	P	G
Charlock	,,	,,	M	G
French Honeysuckle	,,	September	G	G
Mustard	,,	August	M	G
Sainfoin	,,	,,	M	G
Thistle	,,	,,	M	G
Vetch	,,	,,	M	G
Blackhead	July	September	M	G
Lime	,,	August	M	G
Meadowsweet	,,	,,	P	M
Saxifrage	,,	,,	M	M
Thyme	,,	,,	M	G
Blackberry	,,	September	G	M
Heather	,,	,,	P	G
Poppy	,,	,,	G	P
Canterbury Bells	August	November	G	M
Devil's Bit—(Scabiosa Succisa)	,,	September	M	M
Ivy	October	December	M	M

INDEX.

(The Figures denote the pages).

Abdomen, 23.
Aconite, 221, 223.
Adulteration of Wax, 35, 61.
After-swarms, 10, 123; prevention of 126, 134.
Air-space between hive walls, 45.
Air vesicles, 21, 25.
Alighting board, 44.
American-cloth sheet, 51.
Anatomy of Bee—External skeleton, head, simple eyes, compound eyes, 15; antennæ, organs of mouth, 17; thorax, 18; legs, 19; feet, 20; wings, spiracles, and tracheæ, 21; abdomen, honey-sac, sting, 23; palpi, queen's sting, organs of drone, 25; organs of queen, 27.
Anemone, 5.
Antennæ, 17.
Ants, 34, 202.
Apiary, arranging an, 80, 159; selecting position for, 80; house, 82.
Appliance press, 82.
Artificial pollen, 111, 182.
Artificial swarming, 126-129, 167, 194.
Asperser, 90.
Aubrietias, 221, 223
Automatic transfer of bees, 138.
Autumn, the bee in, 13; feeding, 180; syrup, 181; general management, 209; flowers, 221-223.

Balling the queen, 165.
Bean, 25.
Bee, classification of, 1; in spring, 4; summer, 7; autumn and winter, 13; anatomy of, 15-29; flight of, 21, 85; diseases, 185-201.
Bee dress for ladies, 69, 97.
Bee escapes, 46, 100, 150-153.
Bee flowers and plants, 221-223.
Bee food, recipes, 183-184c.
Bee gloves, 70, 71; use of, 70, 96, 130.
Bee-keeping, ancient, 76; modern, 76; delightful occupation, 77; profitable, 79; commencing, 83-91.
Bee literature, 90, 91, 209.
Bee master, what constitutes a, 93.
Bee moth, 203.
Bee pest (foul brood), 188-200.
Bee products, 33-39, 200, exhibiting and judging, 210-220.
Bee space, 43, 51, 53, 54, 55, 145.

Bee stings, 24, 25, 93, 95, 96; treatment of, 97; and rheumatism, 98; protection, 96, 130.
Bee veils, 69, 96, 130.
Bees, natural history of, 1-14; cleanliness of, 4, 6; as fertilisers, 5, 33, 38; unselfishness of, 7; different races of, 30-32; Black or Natives, Italians or Ligurians, Carniolans, 30, 102; Caucasians, 32; Cyprians, 31, 102; Syrians, 31, 102, 113; Giant, Common East-Indian Dwarf East-Indian, 31; Dutch, Sand, Leaf-cutter, 32; feeding, 64, 178-184c; subduing and handling, 67, virtues and beauties of, 76; near dwellings, 80; purchasing 83; value of, 83; moving, 84-87; weight of, 85; driven, 88; driving, 88; uniting, 90, 135-136; fearlessness of, 93; swarming harmless, 94; removing from combs, 105; metamorphosis of, 117; hiving, 130-134; transferring, 136-138; robbing and fighting, 175-177; diseases of, 185-200; enemies of, 202-204; wintering, 205-207; general management of, 208-209. See also queen, workers, drones.
Beeswax, see Wax.
Birds as enemies of bees, 202.
Bi-sulphite of carbon, 194, 202.
Black, or Native Bees, 30, 102.
Blackberry, 14, 223
Black brood, 187.
Blackhead, 221, 223.
Blind Louse, 203.
Borage, 221, 223.
Box, 221, 223.
Braula cœca, 203.
Breeding, begins, 4, 108; stimulating in spring, 110, 139; stimulating in autumn, 117; nursing, 109, 112, 114, 115; controlling drone rearing, 113, 125; limiting queen rearing, 126; ceases, 117; metamorphosis, 117.
Brood, worker, 109; spreading the, 111, 139; drone, 112; queen, 113, 115.
Broom, 221, 223.

Cages for queens, 128, 134, 166.
Candy, soft, 184, 205; flour, 184b, 206.
Cane sugar, 33, 179, 183.
Canterbury bells, 221, 223.

INDEX.

Cappings of cells, 34, 37, 190, 198.
Carbolic acid, 67, 100, 130; cloth, 67, 95, 99, 100, 201; feather, 101, 110.
Carniolan bees, 30, 102.
Casts, 10, 123; preventing, 126, 134.
Caucasian bees, 32.
Cells, worker, 35; drone, 35, 104, 112; hexagonal, 35; transitional, 36; queen, 37, 105, 113; use of for storing, 37; cappings of, 34, 37; number in comb of standard frame, 108; foul brood, 189.
Chaff, between hive walls, 45; cushion, 51. 206.
Cheshire, 62, 187, 190, 197.
Chilled brood, 187.
Classification of honey bee, 1.
"Claustral" Detention Chamber, 52.
Claws, 20.
Cleaning and disinfecting hives, 136, 188, 192, 194, 196, 201, 208.
Clover, 7, 221-223.
Colon, 23.
Colour of hives, 81, 159.
Comb, building, 5, 37; described, 35, 104; value of, 35, 37, 61; stand, 99; box, 99, 100; repaired, 138; feeding for building, 181.
Comb-foundation (see Foundation).
Combs, melting into wax, 72-75, 157, 215-216; removing bees from, 105; turning, 106; cleaning extracted, 152, 156.
Condemned bees, 88.
Cone escape, 46, 100, 150, 177.
Corbiculæ, 5, 19, 38.
Cork Dust, 45, 206.
Cotoneaster, 221, 223.
Crate, 54; divisional, 55, 148; observatory, 56; hanging, 56, 145, 148.
Crates, storing, 139; preparing, 140; putting on, 143; tiering, 146; removing, 147, 150-153; travelling, 171.
Cream of tartar in bee-food, 184.
Crocus, 5, 111, 221, 223.
Crystallised honey, 169, 173, 174, 210.
Currant, 6.
Cyanide of potassium, 203.
Cyprian bees, 31, 102.

Damp roofs, 46, 207.
Dandelion, 5, 223.
Dead bees, removal of, 206, 208, 209.
Differential Diagnosis, 201.
Digestive system, 22.
Diseases, 185; Examining for, 84; dysentery, 185; paralysis, 186; chilled brood, 187; black brood, 187; pickled brood, 188; foul brood, 188-198; "Isle of Wight Disease," 185, 198; differential diagnosis, 201; recipes for treatment, 201.
Division board, 49-50.
Dorsal plates, 22.

Double rabbet, 45.
Double walls, 45.
Doubling, 148.
Dress for lady bee-keepers, 97.
Driving, bees, 88; irons, 89; box, 89.
Drone-breeding queens, 109.
Drone cells, 35, 104, 112.
Drones, described, 3; their brief life, 3; death, 3, 13, 27; eyes, 11; tongue, 18; wings, 21; fertility, 12, 27; organs, 25; dwarf, 109.
Ductus ejaculatorius, 25.
Dummy, 49-50.
Dutch bees, 32.
Dysentery, 185, 206.
Dzierzon, 28, 42.

Earwigs in hives, 202.
East-Indian bees, common, 31; dwarf, 31.
Eggs, fertilisation of, 28; age of, 105.
Enemies of bees, 202.
Entrances, 45; reducing, 176, 179; examining, 206, 208, 209; enlarging, 125, 208.
Epipharynx, 17.
Examinations for experts' certificates, 70.
Examining combs, 103; stocks, 159, 208; roofs, 209.
Excluder, 57; use of, 145, 148.
Exhibiting bee products, 210-218.
Extracting, honey, 154, 214-215; wax, 157, 216.
Extractors, honey, 72; wax, 74.
Eyes of bees, simple and compound, 15; drones, 11, 15; workers, 15; queens, 15.

"Federation" hive, 44; dummy, 50; bee-escape, 152.
Feeders, 64-66.
Feeding, 64, 178; spring stimulative, 110, 179; artificial pollen, 111; objects of, 178; precautions, 179; summer, 180; autumn, 180; quantity of food required, 180; winter, 181, 205; for comb building, 181; in skeps, 182; Recipes, 183-184c.
Feet, 20.
Fertilisation, of egg, 28; of queen, 2, 11, 25, 28.
Fertility, of queen, 2, 4, 12, 28, 29, 108, 109; of drone, 12, 27.
Fighting, 175.
Floor-board, 44; moveable, a *sine qua non*, 192.
Flowers and plants, 221-223.
Follower, 56, 140-141.
Food, see Feeding.
Formalin, 193; treatment for foul brood, 192, 201.
Foul brood, 188-198.
Foundation, fixing in frames, 52, 61, 143; in sections, 54, 140-142; use

of 58, 113, 125; invention of, 59; varieties of, 59; advantages of, 60; adulteration of, 61; testing, 61; change of colour, 61; quantity, required, 61; wiring, 62, 143.
Frames, standard, 51; various, 52; lifting, 99; turning, 106; preparing, 142; wiring, 142.
Fruit trees, 221-223.

Galleria cereana, 203.
General management, work for the month, 208-209.
Giant bees, 31.
Glands, poison, 25; wax, 35.
Gloves, use of, 70, 96, 130; various, 71.
Gooseberry, 221, 223.
Gorse, 5, 221, 223.
Granulation of honey, 169, 173, 174.

Hazel, 5, 223.
Head, 15.
Heather, 14, 221-223; honey, 156, 210, 212, 215, 221, 222.
Heddon methods, 134, 138b.
Hellebore, 221, 223.
Hive, occupants of, 1; sanitation in, 6.
Hives ancient, 40; Skep, 41; moveable comb, 41-43; internal measurements, 43; timber used in, 44; "Federation," 44; floorboard, 44; ventilator, 44, 46; body box, 45; lift, 45; roof, 46, 207; W.B.C., 46; Observatory, 47; I.B.A., 1909," 48; "Hibernian," 49; position of, in apiary, 81, 159; colour of, 81, 159; levelling, 81, 130; stands, 81; nucleus, 159; cleaning and disinfecting 194, 196, 201.
Hiving bees, 130-134; board, 130, 133, 135, 136, 195.
Honey, described, 33, 221-222; gathering and storing, 18, 33; water in, 34; as food, 34; quantity used in wax production, 35, 37; adulteration of, 39; surplus, 139; extracted more profitable than comb, 139; extracting, 72, 154, 214, 215; straining and ripening, 156, 214; marketing, 169-174; Home, 169; imports, 169; storing, 169; grading, 170, 212; bottling, 173, 214-215; packing, 174; crystallised, 169, 173, 174; for exhibition, 210-216.
Honey-Comb (see Comb).
Honey Dew, 34.
Honey extractor, 72, 154; invention of, 72; uncapping knife, 73, 154; strainer and ripener, 73.
Honey flow, 139, 143.
Honey jars, 173.
Honey labels, 174.
Honey press, 74, 156.
Honey sac, 23, 33.
Honey tins, 174.

Huber, 41, 62, 77, 115, 116, 166.

Ileum, 23.
In-breeding avoided in nature, 120.
Intestines, 23.
Introducing queens, 165-168.
IRISH BEE JOURNAL, 91, 209.
Irish Bee-Keepers' Association, examinations, 70.
"Isle of Wight Disease," 185, 198.
Italian bees, 30, 102.
Ivy, 14, 222, 223.
Izal, 130, 183-184b, 192, 201.

Jars for honey, 173, 215.
Judging bee products, 218.

Labels for honey, etc., 174, 215 218.
Labial palpi, 18.
Labium, 17.
Labrum, 17.
Langstroth, 42, 59, 72, 95.
Larvæ, age of, 109.
Laying workers, 12, 31, 115; removing, 116.
Leafcutter bees, 32.
Legs, 19.
Leuckart, 114.
Lifting frames, 99.
Ligurian bees, 30, 102.
Limanthes Douglasii, 221, 223.
Lime, 221, 223.
Lingua, 18.
Loss of Queens, 158.
Lubbock, 34.

Maeterlinck, 95, 166.
Maggot (stylops), 203.
Mandibles, 17.
Manipulating, 99.
Marketing honey, 169-174.
Maxillæ, 17.
Maxillary palpi, 17.
Mead, for exhibition, 217.
Measures, 184c.
Medicated bee-food, 183-184b, 188, 192.
Mehring, 59, 72.
Mentum, 18.
Mice, 202.
Mignonette, 221, 223.
Modern bee-keeping, 76, 77.
Month, Work for the, 208-209.
Mouth, organs, 17.
Moveable-comb hives (See Hives).
Moving bees, 84, 85.
Mysterious Influence, The, 7.

Naphthaline, 191, 202.
Naphthol-Beta solution, 183-184b, 192.
Natural history of bees, 1-14.
Natural swarming, 7-10, 118-126.
Nectar, 33; how gathered, 3, 5, 18, 33.
Nuclei, 162-165.
Nucleus hives, 159.

Œsophagus, 23.
Ovaries of queen and workers, 27, 28.
Oviducts of queen, 27, 28.

Packing bees for transport, 84-87.
Packing honey—comb, 171; extracted, 173.
Palpi, 17, 18, 25.
Paraglossæ, 18.
Paralysis, 186.
Parasites, 203.
Parthenogenesis, 11, 28.
Past and present, 76.
Pea flour, 111, 184b.
Pear, 223.
Petiole, 23.
Petroleum jelly, 99, 100, 143.
Pickled brood, 188.
Pine, seasoned, for hives, 44.
Pipe cover, queen cage, 167.
Planta, 19.
Pliny, 40.
Pollen, 3, 5, 33, 38; how gathered, 19, 38; artificial, 111, 182.
Pollen mite, 203.
Poppy, 221, 223.
Porter bee escape, 152, 177.
Potato masher, 74, 156.
Prevention of Swarming, 124-126.
Profitable industry Bee-keeping a, 79.
Propolis, 3, 39; how gathered, 19, 39.
Pulvillus, 20.
Purchasing bees, 83.
Pyrus japonica, 221, 223.

Queen, described, 2; length of life, 2, 12, 115; her fertilisation, 2, 25, 28; egg-laying powers, 2, 4, 12, 28, 29, 108, 109; virgin queen, 10, 122, 123; her wedding, 11, 81; queen's tongue, 18; sting, 25; organs, 27; finding the queen, 103; searching for the, 107; drone-breeding, 109; caging, 134, 136; loss of, 158; old, 158; defective, 158; "balled," 158, 165; clipping her wings, 121.
Queen cages, 128, 134, 166.
Queen cells, 8, 37, 105, 113; false, 113; inserting, 162.
Queen excluder, 57, 145, 148, 149.
Queen introduction, 165-168.
Queen larvæ, nursing, 114.
Queen rearing, 8, 159-165; limiting, 127; on a large scale, 168.
Queenlessness, 158; signs of, 159.
Queens per post, sending, 168.
Quilts, 50.

Rabbet, Double, 45.
Ragweed, 221, 223.
Reaumur, 77, 203.
Recipes, for bee-food, 183; carbolic solutions, 201; formalin solutions, 201.

Remedies for bee stings, 97.
Re-queening, 196. See queen rearing.
Ribes Rubra, 221, 223.
Ripener, honey, 73, 156.
Ripening honey, 73, 156.
Robbing, 64, 154, 175; precautions, 175; signs of, 175; treatment, 176.
Roofs of hives, 46; defective, to repair, 207, 208, 209.
Root, A. I., 60.
Royal jelly, 114.

Sainfoin, 221, 223.
Sally, 221, 223.
Sand bees, 32.
Saw-dust, 45.
Saxifrage, 221, 223.
Scent diffuser, 90.
Scilla, 221, 223.
Second, third, and fourth casts, 117; 123.
Section, the, 53.
Section crate. See Crate
Section honey less profitable than extracted honey, 139.
Section travelling crate, 171.
Sections, various makes, 53, 140-141; "bait," 56, 145, 156; preparing, 140, 211-213; extracting honey from, 156; cleaning, 156; storing, 156; grading, 170, 212; glazing, 170, 213; packing, 171; for exhibition, 210-214.
Seminal vesicles, 25.
Separators, 54, 140-141.
Shallow frames, 52.
Sheet and quilts, 50.
Silex, 5.
Simmins, 95, 168.
Skeleton of bee, external, 15.
Skep, the, 41; use of, 41, 131-133; automatic transfer from, 138; giving place to moveable-comb hive, 41; supering, 150; feeding bees in, 182; protecting in winter, 207.
Smoker, the, 67, 95, 99; preparing, 100.
Smoking overdone, 102.
Snow, to be removed, 207, 209.
Solar wax extractor, 74, 157.
Spermatheca, queen's, 12, 27, 28.
Spermatophore, 25, 27.
Spermatozoa, 12, 25, 27, 28.
Spiracles, 21.
Spoon, 18.
Spreading the brood, 111, 139.
Spring, the bee in, 4-6; cleaning, 136; feeding, 179, 183; management, 208; flowers, 221-223.
Standard frame, 51; number of cells in, 108.
Stand for combs, 99.
Steam wax extractor, 75, 157.
Stimulative feeding, 110, 117, 161, 179, 208, 209.
Sting, worker's, 24; queen's, 25; use of, 25, 93, 95; unprovoked use of

exceptional, 93; effect of, 96; treatment of, 97; cure for rheumatism, 98; protection from, 96, 130.
Stock, defined, 83; commencing with a, 85.
Stocks, value of, 83; moving, 85; uniting, 135.
Stomach, chyle, 23; mouth, 23, 33.
Storifying, 148.
Storms, precautions against, 207.
Strainer and Ripener, 73, 156.
Study the subject, 90.
Stylops, 203.
Subduing bees, 92.
Sulphur, 194, 204.
Summer, the bee in, 7-12; feeding, 180; general management, 208-209; flowers, 221-223.
Super-box, 57; putting on, 145, removing, 150-153.
Super-clearers, 151-153, 156 211-212.
Supering, 53, 143-150, 211-212; skeps, 150.
Supers, removing, 147, 150-153; for exhibition, 216.
Surplus honey, 139-153.
Survival of the unfit, 76.
Swammerdam, 77.
Swarm, 7-10, 119; defined, 83; commencing with a, 84; value of a, 83, 85, 118; catcher, 132.
Swarming, natural, 8, 118, 119; signs of, 118; delay of, 119; to encourage clustering, 120; prevention of, 124, 126, 134; "fever," 124; artificial, 126-129, 167, 194.
Swarms moving, 84; weighing, 85; sending per post, 85; vagaries of, 119; truant, 120; casts, 10, 123, 126, 134; hunger, 123, 175; making for sale, 128; hiving, 130-134; feeding, 131, 133; retracing, 134; separating, 138b.
Sycamore, 6, 221, 223.
Syrian bees, 31, 102, 113.
Syrup for feeding, recipes, 183, 184c; scented, 90, 135.

Tarsus, 20.
Testes, 25.
Thorax, 18.
Tibia, 19.
Tin bars for section crates, 55, 141.
Tins for extracted honey, 174.
Tongue, 18.
Trachea, 11, 21, 25.

Transferring bees, 136; from skep or box, to modern hive, 137; automatic transfer, 138; Heddon method, 138b.
Uncapping knife, 73.
Uniting bees, 90, 135-136.

Value of combs, 35, 37, 61; of stocks, 83; of swarms, 83, 85, 118.
Vasa deferentia, 25.
Vaseline, 99, 100, 137, 143, 145.
Veils, use of, 69, 96, 130; Lady's, 69 97; wire-cloth, 69.
Velum, 19.
Ventilation, 125, 206.
Ventilators, 44, 46, 151, 193, 197, 198.
Ventral plates, 23.
Vesiculæ seminales, 25.
Vinegar, in bee food, 183; for exhibition, 218.
Virgil, 40, 76.

Wasps, to destroy, 203.
Water for bees, 5, 182, 208.
Water in honey, 34, 182.
Waterproof, making roofs, 207.
Wax, described, 34; production, 5, 19, 23, 34; honey used in, 35, 37; paraffin and ceresin, 35, 61; adulteration, 35, 61; testing for adulteration, 61; extraction, 74, 157, 216; extractors, 74; solar, 74, 157; steam, 75, 157; for exhibition, 216.
Wax moth, 203.
Weed, E. B., 60.
Wheat flour for bee food, 184b.
White Clover, 7, 221, 223.
White rock, 221, 223.
White thorn, 6, 221, 223.
Willow, 221, 223.
Wings, 21.
Winter, the bee in, 13-14; food, 181 183; passages, 205; general management, 208-209; flowers, 221-223.
Wintering, 205-209.
Wiring appliances, 62, 142-143; frames 142.
Work for the month, 208-209.
Worker bees, described, 2; their brief life, 2; work, 2, 5; anatomy, 15-25.
Worker cells, 35; size, 35; number in comb of standard frame, 108.
Yadil, 130, 183, 184, 184b.
Zinc bars for section crates, 55, 141.
Zinc excluder, 57; use of, 145, 148, 149.

TONE BLOCK ILLUSTRATIONS.

Author, Frontispiece.
Queen, Worker, and Drone, 2.
Queen Cells, 9, 163.
Bee on Clover, 33.
Hives, 41, 42, 43, 47, 48, 49, 160, 207.
Appliances, 45, 49, 50, 54, 55, 56, 57, 63, 65, 66, 67, 69, 70, 75, 81, 82, 85, 89, 90, 99, 100, 151, 152, 167, 168, 170, 171, 172, 173, 174, 195, 198, 204, 213.
Claustral Hives, 52B.
Foundation, 59, 60, 63.
I.B.A. Bee Tent, 68.
Examining the Top Crate, 71.
Extractors, 72, 73, 74, 75.
Outbreak of Bee Fever, 78.
Hives on Flags, facing page 80.
Mr. G. Skevington's Apiary, facing page 92.
Subdued Bees on Combs, 92, 94, 108, 164.
Bee Dress for Ladies, 97.
Using the Smoker, 98.
Miss W. Seadon Driving Bees, 99.

Drawing on Carbolic Cloth, 101.
"Thumping" Bees off a Comb, 105.
Clipping Queen's Wing, 121.
Congestion. Bees Crowded out, 129.
Hiving Bees, 131.
Swarm in a High Tree, 132.
Repaired Comb, 138.
Rescuing Condemned Bees, 138c.
Fixing Foundation in Sections, 140.
Putting on a Crate, 144.
   ,, ,, Super Box, 145.
Removing a Crate, 147.
Crates Tiered, 148.
Hive Doubled, 149.
Supering a Skep, 150.
Removing a Frame Super, 151.
Extracting Honey, 155.
Queen Rearing, 164; Appliances, 168b.
Comb infected by Foul Brood, 189.
Diseased Stock, "J.G.D." Ventilator, &c., 197, 198.
"Isle of Wight Disease," 199.
Water Fountain, 204.
Thatched Skep, 206.
Hives prepared for Summer and Winter, 207.

THE C.D.B. HIVE.

EDMONDSON BROTHERS.

BEE HIVE MANUFACTURERS.

WINNER OF 10 FIRST PRIZE MEDALS.

This Hive was selected by the Congested Districts Board for supply to the Beekeepers in Donegal and West and South of Ireland, and is the best turned-out Hive in the British Islands. The internal fittings are eleven standard bar frames, dummy, three quilts, and one crate of sections. The ventilators are fitted in front with a pair of Cone Escapes, and at back protected with perforated zinc.

"Two Crate" and "Cottagers" Hives. Sections, Foundations, Frames, &c.

10 DAME ST., DUBLIN